*Your interest.
and interview.
Hank C[...]*

THE ILLUSION OF KNOWLEDGE

The Illusion of Knowledge

Copyright © 2021 by Harold Katcher
All rights reserved
The moral rights of the author have been asserted

Cover image by © khak - stock.adobe.com

International Cataloging Data in Publication (CIP)
Angelica Ilacqua CRB-8/7057

Katcher, Harold
   The illusion of knowledge : the paradigm shift in aging research that shows the way to human rejuvenation / Harold Katcher. -- Valinhos, SP : NTZ , 2021.
   226 p.

ISBN 978-85-54106-05-8

1. Biology 2. Celular aging 3. Rejuvenation I. Title

21-3342   CDD 611.018

Indexes for systematic catalog:

1. Biology

Published by NTZ
www.ntzplural.com
Valinhos/SP - Brazil

First edition

THE ILLUSION OF KNOWLEDGE

The paradigm shift in aging research
that shows the way to human rejuvenation

Harold Katcher, PhD

NTZ

2021

Table of Contents

Foreword..7
1. The birth of death..13
2. What is "life"?..35
3. Some see "the light"..51
4. Biology begins...73
5. The origin of species...97
6. Why do aging cells die?..129
7. An autobiographical note or two..151
8. The invention and discovery of E5...171
9. The new science of rejuvenation...199
References..213

Foreword

> "The greatest enemy of knowledge is not ignorance; it is the illusion of knowledge."
>
> **Stephen Hawking**

The recent discovery by our group[1] of the ability to revert old rats to young convinces us that mammalian aging — and that means human aging as well — can be reversed, in a process known as "rejuvenation" (literally, "re-youthening", a return to youth). I see this as the next step in human evolution, and like most human evolution, it involves an increase in our powers — not through biological evolution, which is far too slow, but through knowledge, translated into technology. And yet, while every old person would like to be young again — smarter, stronger, more virile or fertile — what might the actual benefit to society be? Having taught *Biology of Aging* for years, I know that the majority of my students don't see rejuvenation as a benefit, but as a burden, because, surely, if you can rejuvenate people, you could add decades to their lives, and then the question that worries us most comes to the fore: overpopulation. Most of my students, as they signed up for a course on the biology of aging, were interested in the subject, and even in anti-aging medicine, but their desire was not for physical immortality and eternal youth (although some had such a desire indeed) — they simply wanted to live what is a long healthy life for about 100 years and die peacefully in their sleep.

At the time of writing this, still relatively near the beginning of the 21st century, the huge "baby boom" that occurred as troops returned home when World War II ended produced a big bulge in populations across the globe. People born during this time were referred to as the "Baby Boomer Generation" or simply "boomers". Those boomers, a large and comparatively wealthy generation of people, have entered into senescence (i.e., old age), and some have realized that the comfort that money and status brought them in their earlier lives meant little when they discovered that the promise of their "golden years" (their reward for a lifetime of hard work and accomplishment) was a lie. Accomplishments and learning meant little except fond (and perhaps disappearing) memories. And why is it a lie? Because now that you have the time, you no longer have the energy. Your "golden years" are a time when mind and body become sources of pain and loss, and the future only bodes worse to come before the end.

Even though the search for immortality has always been a preoccupation of those that have everything else (for it is commonly known that "you can't take it with you"), our comparative wealth nowadays created a different picture, as even the poor in most countries have entertainment (via cell phones, TVs, etc.) that Emperor Alexander couldn't imagine, and in the richest countries even the poor often have air-conditioning and the ability to make trips that would have taken Alexander weeks or months in mere hours. These increases in our wealth and abilities are the result of our increasing knowledge — our science.

The widespread interest by the now aging boomers has led to the creation of major "blockbuster" drugs produced by the pharmaceutical industry — such as statins, metformin and ACE-inhibitors — for treating the diseases of aging, but which require life-long use. And as "aging" itself is not considered a "disease" by the Food and Drug Administration (FDA) of the USA (and by the equivalent government bodies around the world), a "supplement" or "nutraceutical" industry outside mainstream medicine was born, based on the "aging theory" du jour. Multi-vitamins, particularly anti-oxidants (some totally valueless when ingested, like SOD, an enzyme that is digested like any other protein), dedicated to anti-aging treatment are not regulated in the USA by the FDA, as these "supplements" are generally regarded as safe.

The pharmaceutical industry produces drugs that do ease or delay the deficits and diseases of aging — drugs based on known science. It has had some accomplishments: the risk of various diseases of aging has been reduced; some cancers, once untreatable, now have high remission rates; statins and ACE inhibitors lower "bad" cholesterol and blood pressure, respectively, preventing or ameliorating heart and artery disease; we have metformin which helps to take care of type 2 diabetes and gives its users a longer-than-average lifespan.

Nevertheless, current science hasn't produced significant increases in human lifespan, though the average lifespan has been increased considerably, as, at the turn of the 20th century, the average lifespan in the USA was close to 40 years and is now close to 80 years. This is because the average lifespan includes all those deaths from bacterial diseases, once our major killers (plague and tuberculosis), which are mostly prevented by proper sanitation rather than antibiotics. However, as I'll discuss later in this book, scientists have been able to extend the lifespan of non-primate mammals by a significant fraction of their lifespan and have been able to increase the lifespan of some invertebrates several-fold.

At this time, perhaps for the first time in history, we have produced medications, though with side-effects, that can ameliorate and extend our lives (even if they mostly ameliorate and extend our senescence, as they're generally employed by people who are in or close to senescence), increasing our average lifespans, but only a bit — yet, life is precious to many, no matter their condition.

The explanations most biologists give for the processes controlling aging and lifespan length are based on years of brilliant theoretical and experimental work during the mid-to-late 20th century. The result of these myriad efforts was a commonsense explanation of aging, a previously mysterious and unaccounted for attribute of life, that could now be understood by anyone: "Things fall apart..."; or in more scientific language: entropy. In the context of aging, entropy means the inevitable accumulation of random (*stochastic* is the "fancy" word for it) damages that gradually overwhelm the ability of our cells to repair them, and our cells age and die in consequence. So, in this analogy, your body is like your automobile: yes, you take good care of it, but there was that stone you didn't see that damaged your transmission, that rust spreading beneath the paint on your door, the constant wearing of pistons and cylinders. Yes, you could replace every warn and defective part if you had the replacements, and the money, but that would just be throwing your money away and you have other needs, and obtaining those new "parts" would be expensive. So, the best you can do is try to prevent damage from happening and repair it when it does, though you know that damage is inevitable.

To extend the analogy, the better your car was built initially (your genetics), and the better it was taken care of (your diet, exercise regime and prudence), the longer it will function. Modern science metaphorically developed some fuel additives, some synthetic lubricants and radiator sealants to keep the old engine from wearing out too quickly, but the prospect of significant extension of life is small, and barely worth the effort to most (myself included, and even Leonard Hayflick, the man who convinced us that cells themselves are mortal — at least when grown in vitro ["in glass", rather than "in vivo" — in animals]).[2] Some brave (or arrogant) souls nonetheless undertook the challenge of "fixing" Nature's defective machinery, to show Nature what human ingenuity could accomplish (I acknowledge Aubrey de Grey for this);[3] their view of aging is still based on what has been called the "Wear and Tear" theories of aging (the random occurrence and accumulation of damage to vital cellular structures that is unrepaired or unrepairable by the body), and they have little to show for their efforts. The reason for this is simple — these theories have something in common with most commonsensical theories about the world (such as "the Sun circles the Earth once a day"): they are wrong.

For the rest of this book, I will reveal the secrets of aging that I have discovered; some of them were hiding in plain view, and some of them our group have uncovered in our own laboratories. We have been able to reverse the aging process in rats such that the experimentally-treated old rats measured (in terms of biological age) less than half the chronological age of their age-matched controls (untreated old animals, the same chronological age as the treated ones) according to the "gold standard" of biological age determination (Steve Horvath's DNAm age "clock" assays developed specifically for the Sprague Dawley rats we used), and more than 30 different "aging biomarkers" (measurable characteristics such as inflammation, strength, cognitive ability and organ health, that change with age). Human rejuvenation is, of course, our ultimate goal. The question is whether humanity is ready for what will be the greatest change in the course of human history, perhaps as fundamental as the discovery of fire. Like fire, I'm sure we'll find new ways to use it, and like fire, it will change our lives for the better.

It is obvious that many human beings already have more life than they know what to do with — frittering away the time given to them with divertissements; "drugs, sex and rock and roll" or whatever the modern equivalent for disposing of excess time is. For many in this world, life is already too burdensome, and they are ready to give it up for any imagined "just cause" — however, it is my belief that human biological immortality will not only lead to a better world, but to better *worlds*. While the speed of light continues to impose limits on our travels, a lifespan of centuries, millennia or more, is needed for humanity to create a pan-galactic civilization.

My approach in this book is to mix biology with both objective and personal history, and the histories and mythologies of past and present, to get a more complete picture of the potential benefits and deficits of human rejuvenation. This is, to some extent, my personal quest for biological immortality (not that I fear death, I consider it the "Great Adventure", and if there's nothing, I won't be disappointed, but I know I've always been a good and faithful servant so I fear nothing). I will also go into some detail about the past and present theories of aging and the underlying biology and biochemistry that supports (and sometimes contradicts) those ideas.

Rejuvenation, once an "impossible dream", a subject related more to magic than science, has been demonstrated several times within the past several years (though never to the extent we have shown recently). We will discuss both the "Why" and "How" of human rejuvenation, because rejuvenation has changed from an impossible dream to a new reality, one that I hope will be realized during the lifetimes of most of the readers of this book (myself included as my dream foretold — a dream which will be described later on this book).

The ancients, no less than us, were concerned with life and death (and surprisingly, had clues to the process of rejuvenation), so let's start with what ancient peoples believed (and many still believe) about life, death and immortality. We will find that some of this ancient wisdom was closer to the mark than the "wisdom" of modern biologists.

1

The birth of death

> "For in much wisdom is much grief,
> And he who increases knowledge increases sorrow."
>
> **Ecclesiastes 1:18**
>
> "Sorry folks, but this knowledge is actually good news."
>
> **Harold Katcher (the author)**

Grief and mourning among mammals and birds have been anecdotally noted many times, but do we have the slight inkling of whether animals know they face aging and death? Small changes in our knowledge make big changes in our behavior — it is speculated that the discovery of the relationship between having sex and procreation nine months later reversed the relative statuses of men and women, when the magical power of the woman to produce a baby was reduced to her being the receptive soil for the male seed (of course, neither is true).

It is unknowable when humans first discovered aging and death as inevitabilities (and it probably occurred many times in many places), but the belief that some sort of afterlife existed (the denial of death as an end of personality) was common to many cultures. Evidence suggests the possibility that from the very beginning, there were societies that feared death as unnatural and random, and others that saw death as part of life. Of the latter, the Pygmy peoples of the African forests regard themselves as parts of the forest itself, and they believe they return to the forest after death. For these people, death is natural — they even purposely forget the names of those who died and never use them again. However, for most of humankind and most of the history of civilization, the denial of death was a source of higher human culture and the basis of religion, because every human is shadowed by the realization of their own death — and the impossibility of believing that they as individuals will die led to the concept of the afterlife and the "soul".

Mesopotamia

The very first story of the quest for immortality — possibly the first story ever written, inscribed on clay tablets, marked with the dots and slashes of the cuneiform writing system of ancient Sumer, the first human civilization ("civilization" specifically refers to *civitas*, or cities) — is called *The Epic of Gilgamesh*. As in many belief systems of today, neither the Sumerians, nor the Akkadians, that later replaced them, believed in personal death, but rather that death represented a transition to a different phase of life — one that took place in the "Netherworld" — a world on the other side of the Earth, where the sun god Shamash (his name in the Akkadian language) spent every night dispensing justice to the dead living there (as that was one of his roles). But the Netherworld was a dark and dreary land where the ghosts of the living ate dust and drank brackish water unless food and drink were provided for by living relatives — there were actually pipes leading into graves where libations could be poured — for the spirit of the deceased still required sustenance and could still interfere in the affairs of the living (negatively, if those spirits weren't properly mollified). From the number of votive figures found, religion was a major industry then.

Gilgamesh, from *The Epic of Gilgamesh*, was the king of wealthy Uruk — a Sumerian city-state in what would be Iraq today — and a fierce and powerful ruler, who angered his people by using his *droit de seigneur** to possess every new bride at her wedding. In response to entreaties from the citizens of Uruk, the Gods fashioned a friend for Gilgamesh, a man his equal in strength, the wild-man Enkidu. Enkidu was lured from the wild with beer and women (specifically, the temple prostitute Shamhat) and brought into Uruk. Eventually, after Enkidu hears of what Gilgamesh does to young brides, Enkidu too appears at a wedding and fights Gilgamesh for the honor of the new bride, and though he loses to Gilgamesh, the two then become inseparable friends, and fierce Gilgamesh changes his demeanor to become a hero to his city.

In their epic adventures together, Enkidu kills the demigod Humbaba and the great Bull of Heaven and is killed by the gods in retribution. Gilgamesh is now alone and sees what death has reduced his friend to, and greatly fears it. Thus, he begins his journey to the island Faraway to find the only immortal man (who lived with his equally immortal wife): Utnapishtim (he is also the model for the man known as Noah in the Hebrew Bible, one of the few to survive the Great Flood). The journey, as can be imagined, was full of wonders (they take the road the sun god uses, under the mountains and past the jewel-bearing trees in the kingdom of the gods), but when Gilgamesh finally meets Utnapishtim, the immortal convinces Gilgamesh that his quest is for naught, that the Gods themselves planted the seed of death in every creation, but offers Gilgamesh a challenge that will give him immortality if met: to stay awake for six days — but he fails.

Once Gilgamesh failed the test, Utnapishtim tells him that he should enjoy the pleasures of life and not waste his time on this useless quest. Gilgamesh is disconsolate, and Utnapishtim's wife, out of pity, makes Utnapishtim tell Gilgamesh about a plant found at the bottom of the ocean that will make him young again. Gilgamesh travels there by tying heavy rocks to his feet, finds the youth plant, and brings it to the surface. Suspicious (and here may be one point of the story), Gilgamesh decides to travel back to Uruk and try the youth-plant out on an old man before using it on himself (an advantage of being king), but on the journey home, while Gilgamesh and his boat captain stop to catch some sleep on an island, a snake eats the

---

* *Droit du seigneur* ("lord's right") was a supposed legal right in medieval Europe, allowing feudal lords to have sexual relations with subordinate women, in particular, on the wedding nights of the women.

youth-plant and from then on rejuvenates itself by shedding its skin each year (a "just so" story) — and Gilgamesh becomes desperate.

For what had he wasted his life? In searching for more life, he wasted the little life he had been given. And yet, after several days of sailing, the great walls of Uruk loom up before them, and Gilgamesh pridefully shows the boat captain his city and his accomplishments, thus leaving Gilgamesh, the great king (two-thirds human and one-third god, but don't ask about the genetics of that) in the same position as we find ourselves.

The ancients, no less than us, were concerned with life and death (and that is the great wisdom in which there is great sorrow), and it appears, in regard to those topics, that nothing much has changed in the past 4,000 years! So, are we still wasting our time searching for more life rather than enjoying what we've been given? Why not accept our destiny and enjoy the life we have? If I were asked that same question even ten years ago, I would say that yes, the *Epic of Gilgamesh* is a "cautionary tale", a sort of "don't throw away what you are given in an unjustifiable expectation of getting more" — but now my opinion has changed (and you'll see why), and I believe that we can have biologically "immortal" youthfulness (with the recognition that nothing physical is immortal and even our universe may have an end).

In the Mesopotamian myths and legends, death did not exist except as a change of state which required a continued burden on the living of supplying the dead (and the priests) food and drink for eternity. Thus, there are now many hungry ghosts in the Netherworld as the dwellers of Uruk, Ur, and Eridu no longer exist to offer them food and drink.

The *Epic* has two hopeful stories to help us accept our fate. Before Enkidu dies, he curses the hunter and the temple prostitute, Shamhat, who lured him from the wild, but after his adventures with Gilgamesh are recounted to him, and the pleasures of civilized life remembered (beer and sex?), Enkidu takes back his curses and praises the two. Though while he was a wild man he knew nothing of aging and death, what he had experienced since then made it all worthwhile. And Gilgamesh, forbidden physical immortality, finds the lesser satisfaction that his name and his deeds will not perish. It's been more than 4,000 years, and yet Gilgamesh and Enkidu still talk to us, so he has achieved that much.

The Hebrews

Judaism

Judaism arose in Mesopotamia as followers of a radical "breaker of idols" in favor of an invisible God. First merely a tribal God, one among many ("Elohim", the word used in the Hebrew Bible that usually refers to a single deity, means *gods* — it has a plural ending "im" ["eem"] that acts as the "s" in English words), their God, Yahweh, became the only God, the one God, the King of the Universe. No longer an amplified man or woman, with lusts and other human desires, this invisible God, whose "thoughts are not like our thoughts and whose doings are beyond what we can imagine", marked a huge departure in the ancient world, from pantheons of concrete gods carved from stone — beings imagined to be giant and powerful extrapolations of men and women — to an invisible God, the only God, a being who alone controls the entire universe (such as was known).

However, another profound distinction separated Judaism and the descendants of Abraham from the similar peoples surrounding them (Abraham was himself the son of an idol-maker from the Chaldean city of Ur): they understood death as the final end of life. An "afterlife" or "Netherworld" did not exist in the Jewish worldview — death was final. Death as the final end of consciousness was born. Only later, in their dealings with the Hellenic Greeks, did some of their metaphysical concepts enter into Judaism such that now even some Jews who consider themselves "orthodox" believe in an afterlife and in immortal souls, but this appears nowhere in the "Old Testament".

One must not neglect the influence in Judaism of the Persian Zoroastrian religion, which presumed an immortal soul and a tripartite God, Ahura Mazda, from whom two spirits arose, much the same as Yin and Yang in Taoism. This was the symbol of two opposites creating the whole; light and dark, spirit and matter, the "increaser" and the "destroyer", and it remained in the Abrahamic faiths as Good and Evil, God and Devil, after Ahura Mazda and Ahriman. Furthermore, as the soul of the dead person is judged and consigned to either a delightful afterlife or a horrible one, based on their deeds and on their righteousness, and not burial ceremonies or bribes to the gods, the concepts of Heaven and Hell, both prominent and even defining for Christian and Muslim faiths, came from the Zoroastrian religion. Additionally, angels and other spiritual beings also came from this still extant though smallest of the world religions.

In the Hebrew Bible, the creation of the world ends with the man Adam being created from red clay (the meaning of "Adam") into which God breathes life. Man, at this point without female representation, was immortal. And God, wishing to give man companionship, created the female out of one of his ribs. They were both immortal[*] and lived in Paradise, and could directly speak to God.[4]

In the first "recorded" (though Adam, being ignorant of writing, didn't record them) words from God to humanity, Adam and Eve were told they could eat from any tree in the Garden, except for the "Tree of Knowledge of Good and Evil" that lay at its center, and that if they did so, they would surely die. Now nearly everyone knows the rest of the tale; a snake convinces Eve to eat from the "Tree of Knowledge of Good and Evil" by telling her that surely she will not die and will have knowledge of good and evil like God. Eve convinces Adam to do the same. And together, eating of the fruit, they lose their innocence and cover their nakedness, thus showing God that they knew good and evil and therefore ate from the tree.[4]

The punishment is quite severe, and they are banished from the Garden, and the Bible then tells us that the "LORD God said: 'The man has now become like one of us, knowing good and evil. He must not be allowed to reach out his hand and take also from the Tree of Life and eat, and live forever'. So the LORD God banished him from the Garden of Eden to work the ground from which he had been taken. After he drove the man out, he placed on the east side of the Garden of Eden cherubim and a flaming sword flashing back and forth to guard the way to the tree of life." (Genesis 2:4-3:24)[4]

Well, it looks like that prohibition has been removed.

There's an awful lot packed into that story, and a lot of questions remain. Why did God put the Tree of Knowledge of Good and Evil at the center of the garden, and why put the tree in the garden at all? And if Adam and Eve had no knowledge of good and evil, why would they think that disobeying God's instructions was wrong? Is it kind of like your dog when you say "sit" and it doesn't? And finally, God does really act the part of a jealous God, somehow fearful that man, knowing what gods know, might eat of the

---

* It suddenly occurs to me that Adam and Eve were not immortal, as God feared Adam and Eve would now eat from the "Tree of Life" and become gods themselves in the myth. As the fruit of the Tree of Life conferred eternal life, and they were barred from eating it, Adam and Eve were mortal.

Tree of Life and gain immortality. And God says that man has become "like one of us", implying that there are more gods (beings who differ from us because they have immortality?).

So, before the "Fall", might we assume that people who have no knowledge of good and evil would be like other animals, for whom "good" and "evil" mean nothing? Not knowing not to disobey God implies knowledge of good and evil (because Adam and Eve are punished for disobeying, but how are they to know disobeying is evil?). However, sexuality does appear to be the concern of "good" and "evil" (they discover their nakedness and dress themselves in fig leaves from shame), and the origin of death. As a biologist, I believe this to be true, as sexual reproduction is the basis of variety, and death by aging is a natural basis to allow for the selection of the fittest, given the natural advantages of size and, in higher mammals, knowledge of the parental generation.

Something else appears in the Jewish Bible (Old Testament) that has particular significance to our interests concerning life and death, and for me, two quotations have particular relevance:

1. "Our days may come to seventy years, or eighty, if our strength endures; yet the best of them are but trouble and sorrow, for they quickly pass, and we fly away."

    Psalms 90:10[4]

2. "You must not eat the blood of any creature, because the life of every creature is its blood; anyone who eats it must be cut off."

    Leviticus 17:14[4]

And finally, a third quote that will illustrate the differences between ancient and modern times:

3. "What has been will be again,
    what has been done will be done again;
    there is nothing new under the sun.
    Is there anything of which one can say,
    'Look! This is something new'?
    It was here already, long ago;
    it was here before our time.
    No one remembers the former generations,"

    Ecclesiastes 1:2-11[4]

Let's discuss these three quotes in relation to our main topics: life, death and rejuvenation.

The first of these quotes comes from the psalms of King David (c. 1,000 BCE), an important figure in the three Abrahamic religions, Judaism, Christianity and Islam ("nice" fellow he was, famous singer and warrior — he had all of King Saul's sons crucified when he became king). What David sang in the very first quote (1) is a prayer by a man named Moses (not "the" Moses), who prays to God and asks for his love and protection — for God will satisfy those who call on him with long life. And here clearly it is stated that you will be given seventy or eighty years of life, depending on your strength, but years full of trouble and sorrow, and here Moses simply asks God to give him as much time in happiness as in affliction.

There is no talk of an afterlife — if Judaism has a motto, it is "Life" (L'Chaim). There is no "soul" in classical Judaism; the word normally translated as "soul" is the Hebrew word *neshamah,* which means "breath" (like the breath God breathed into Adam, who he made from red clay and brought to life as the first man). This sort of soul did not exist apart from the body. That was the definition of death — lack of breathing; without breath, no life. In that case, we ought to be able to say "The breath of a person is their life", but we cannot — as we'll consider when we examine our next Biblical quotation.

In summary, thus far, what I take as our message from the Judaism of the "Old Testament" is that lifespan is fixed, given a small range of potential lifespan durations (70 - 80 years), and death is permanent.

Judaism survives today as an ethnic religion, a "family religion" in ways like Hinduism — even Christian Indians know their caste. You are born a Hindu as were your parents and their parents, and so on. However, unlike Hinduism, there is an official ceremony for conversion to Judaism (indeed the "Book of Ruth" concerns a celebrated woman who left her people to join her Jewish husband's people) though the present-day religion is divided into Reform, Conservative and "Orthodox Judaism", which is hardly orthodox in the Biblical sense; instead it is based on the teachings of the mystic Baal Shem Tov on trying to merge with God in this life (one way of avoiding death), very much like the branch of mystical Islam called Sufism based on the writings of the Persian mystic and mullah Rumi (once, the most widely read poet in the USA).

This ability to see a world beyond our world and greater than it (greater than words can express), to see beauty in the world others cannot, is a splendid answer to death, but is it true or self-delusion? And does it even matter? With dietary restriction plus life-extending mutations, a roundworm

(*Caenorhabditis elegans*) can live ten times as long. Do you imagine the worm appreciates this? On the other hand, if you're old and could have back the body (and brain) of a twenty-something, wouldn't you want to try it? Unlike those popular "Netflix" dramas where people are "cursed" with immortality, there will be no permanent "cure" — aging will start again after treatment, but should only require another round of treatment (we're talking five to ten years intervals, possibly longer), so aging can always be reversed.

The second quote is God telling us "the blood is the life" ("...the life of every creature is in its blood"). That is a phrase I first encountered not in a Bible, but watching a horror movie. This is what "Count Dracula" says in the Bram Stoker novel *Dracula* and is said dramatically by Bela Lugosi in the most famous of the more than 170 films (worldwide) based on the Dracula novel. The underlying thesis of the vampire myth is the Biblical verse (2) quoted above. The vampire artificially sustains its life and youthful appearance by draining the life (blood) from young people. The victims are usually young people (in Stoker's novel, the Count's wives are particularly pleased to be tossed an infant by their master). Does this perhaps hint that a young person had more "life"?

That was believed by the "real vampire" on whom Dracula may have been based: the Hungarian noblewoman Elizabeth Bathory (often referred to as the "Vampire Queen"), who also apparently believed that drinking (and reputedly bathing in) the blood of young women helped her retain her youth. I haven't heard evidence that she did in fact retain her youth (hundreds of young women are said to have died for that cause), but her love of torture and sexual mutilation suggests other motives as well.

What particularly surprises me though is the truth of that Biblical phrase, "the blood is the life of every creature" — if properly understood. If the breath is the animating force, the *neshamah*, the "breath" that brought the clay that was Adam to life, then why do we have God Himself saying "the life of any creature is in its blood"? When I heard that, I thought about it differently — was there any truth to that statement of "life" being in the blood? Why even consider this ancient text? And yet, as we will come to find out, there's great truth in that statement.

The third and final quote from the Jewish Bible is the words of the man referred to only as "the Preacher", or "the Teacher". And the reason I'm quoting this is because this is a philosophy that ruled the world up until the Enlightenment in Europe during the 18th and 19th centuries, and is still believed by many: time is cyclic, there is nothing new under the sun — what was done, has been done before, and will be done again, though none remember, time after time, endlessly.

It doesn't take much serious, evidence-based discussion to arrive at today's truth, a truth that belies the Teacher's words. Never before have men and their machines stood on the surface of the Moon, explored Mars, mapped the depths of the seas, the distances to the stars and the size of atoms; and yet, 3,000 years after David sang Psalm 90, and the Teacher taught his despair, we can but stay death a little, so that lifespan of 70-80 years David sung of has perhaps lengthened five to ten years over the past 30 centuries, but that is about to change.

Early Christianity

This religion started as an offshoot of Judaism, and its doctrines were originally preached by Jesus of Nazareth during the Roman occupation of Israel. In one way at least, Jesus joined the world of the Jews with that of the surrounding Mesopotamian and Egyptian civilizations: death was not eternal, at least for some. The utter despair of death, and the alternative of a short life filled with suffering, was unsupportable to the Jewish people; if they were indeed a "chosen" people, then why were they conquered? Why did they suffer, and why did God bring one into life only to endure pain?

So, to give a reward for good behavior, other than just obeying the Law, and yet keeping the assumption that humans had no immortal souls, Paradise was created — like the Garden of Eden, a place where God lives with his people for all eternity. But instead of an "immortal" soul, Jesus told that there would be a "resurrection" of the dead, who, if they were of God's people, would spend eternity in Paradise with Him. But that place was not in the Heavens, but on Earth, and the "souls" of the dead would not rise, because as discussed, there was no soul apart from the body, but God would raise the dead with new and perfect bodies. He would bring back all the dead, good and bad, but the bad would be brought back merely to finally understand and regret the error of their ways and then be utterly destroyed, like chaff thrown into the fire — so the eternal punishment was not Hell, but simply non-being, eternal death.

However, the dead were dead; they had no souls and were destined to return to the dust from which they came. No souls went to Paradise; instead,

Paradise would come to Earth, but only long after death (though I suppose to the newly recreated, no time would have passed), when God would join Heaven and Earth to create a new Heaven and a new Earth, a perfect world. The resurrected good dead would have eternal lives in perfect bodies, to live with God, and the evil dead would only be resurrected to understand and regret their failings and then obliterated. This concept of resurrection, rather than immortal souls, made early Christianity very different from other religions of the area.

What Jesus did was to change the requirements for entry into Paradise; no longer were strictly following of rules and performing the rituals sufficient for God to find you on "His" side; instead, there were two Biblical commands that to Jesus outweighed and defined all others: "you must love God with all your heart", and "you must love your neighbor". But there was no "immortal soul"; the "soul" did not go anywhere after death — the soul was the breath, when the breath stopped there was no life, no soul. When Paradise comes to Earth, God will judge, keeping the worthy to live with Him in a perfect world and eliminate the bad, those opposed to God, but no eternal tortures such as medieval preachers enjoyed imagining, "just" obliteration and oblivion.

The Greeks

The early Greek religion had an elaborate afterlife, with judgment of the dead and different abodes for the great (Elysium and the Isles of the Blessed), the good (Asphodel Meadows for the ordinary folks), those who wasted their lives on unrequited love (Mourning Fields), and the bad (who were punished in Tartarus, the Underworld's underworld). Tartarus was as far below the Underworld as the sea from the sky, the deepest darkest root of both land and sea. Now, if you were *really* good, demigod level, you might find yourself in the Elysium (or Elysian Fields), or going back to Earth to perfect your soul; once you reach Elysium you are allowed to stay or to reincarnate; if in three incarnations you make your way to Elysium, you spend eternity in the paradise of the Isles of the Blessed (in later Greek thought the Elysium fields were expanded to include ordinary mortals who led virtuous lives, so the Isles of Blessed became the sole abode of demigods and heroes).

However, the soul, for most of the dead, was inactive, a sort of memory that they once existed, with no aims and no consequences to their actions. Death was seen as an unacceptable end to life such that Homer believed that the best possible existence for humans was to never be born at all, or die soon after birth, because the greatness of life could never balance the price of death (interestingly, avoiding Hades' realm was permissible for very young babies, were they to die, implying that the immortal soul does not "enter" the baby until sometime after birth to Homer's way of thinking).[5]

Homer's point was exactly opposite to Enkidu's revelation that the life he lived was worth his death (and knowledge of aging and death he hadn't had as a "wild" man) and Gilgamesh's realization, when he saw the great walls of his city and recounted his deeds, that his earthly accomplishments were worth the knowledge of inevitable death; for Homer, they were not. The feeling was hopelessness, that no matter what you accomplish in life, at least for the common folk (and not the demigod or hero), it all amounts to nothing after you are dead.

One obvious solution already mentioned was to expand the Elysian Fields to common (but exemplary) people, to give people hope, but that hope was better met by Christianity. In fact, by the end of the archaic period, before Christianity took hold, few Greeks believed in the ancient views of the afterlife — though by the Hellenic period most Greeks would put coins in the mouth of the dead for Charon, but this was probably due to the widespread superstition that the spirits of the dead, especially if unburied or if they remained this side of the River Styx by not having a coin for Charon, might return to the living world as ghosts and seek to revenge on people or even whole cities they perceived as having harmed them or simply having more than them.

People used herbs and amulets to fend off ghosts, and in some cases used magical amulets to call forth ghosts to use against other people. Ghosts mostly don't like being called, so many tried to call the ghosts of those who lived a short life (perhaps they wanted to see more of life?). But one Greek innovation that provided material for later religions were mystery cults that would assure their members a place in the Asphodel Meadows to spend eternity in open fields singing and dancing, while non-members would grovel in the mire. Even at the time, the weakness of that solution was apparent and remarked upon; while great people groveled in the mire, not belonging to the cult, much inferior people would enjoy all the pleasures of paradise. Of course, the same argument applies to any religion that only allows its members into whatever is their paradise.

The idea of the immortal soul was already present in Plato's philosophy — in fact, there are modern mathematicians who believe in Plato's abstract "Realm of Forms", and that rather than creating mathematics, they are entering the Realm to discover the mathematics already there (what we know as a "circle" is an inferior depiction of the "true" circle in the Realm). In the same way, the soul existed in the Realm of Forms before our birth and will continue to exist beyond our deaths, forever. Once we die, "our" soul is released and is immortal, and after a brief rest in the Realm, it will enter another body.

It was the fusion of the religion of the Hebrews, the Zoroastrian Persians and Hellenic Greeks that produced the majority of the presently accepted incarnation of the Christian religion, with the new Greek adherents rejecting Jesus' claim of the resurrection of physical bodies (according to the Hebrew idea that the soul [or "breath"] was mortal and dissipated like smoke with the body's death), and substituting it for the more sophisticated "immortal" soul which leaves the body after death, as it never dies, and then enters either everlasting torture or everlasting bliss.

Plato wrote at a time when the popular opinion among Greeks was the same as the Hebrews', that the soul does not survive death, but dies with the body. Plato believed, though I don't see the justification for it, that whatever is self-animating is immortal and the soul is self-animating. He believed that the soul is a non-tangible but understandable "form" like "beauty" and "justice", and like them, it is indestructible (are "beauty" and "justice" self-animating?).

Certainly, what the Greek intelligentsia did to the simplistic Hebrew notion of death being the end of life made it a far more acceptable and understandable answer to life-after-death than "resurrection" (What would be "resurrected"? Would there have to be bones [or DNA] to build upon?). Besides, it was the prevailing view in the ancient world, "perfected" by (rightly) world-famous Greek intellect.

However, I don't understand by which "authority" a view that contradicts what Jesus (who is given his authority by God — at least I would imagine any believer would agree to that) says about death allows the Church to substitute it for the beliefs of Greek philosophers about immortal souls. Let's remember this about the Ancient Greeks: their great philosophers, Plato and Aristotle, formed the opinions of the Catholic Church on secular matters for nearly fifteen hundred years, as the ultimate authorities on the physical universe, but as wonderful as their thinking was, it still harked back to the Greek concept that the world could be understood through pure thought; it cannot.

The Romans

The Roman Underworld, as described in Virgil's *Aeneid*, is a duplicate of the Greeks', with Pluto replacing, in name, Hades, but also with a change in his reputation, with Hades — disliked by gods and men and only prayed to bring death to an enemy — transformed in Roman times in Pluto (God of the Underworld), who had a more positive image. Proserpina (the Roman equivalent of Persephone, the Queen of the Underworld) and Mercury (the Roman equivalent of the Greek Hermes — like Hermes, he guided the souls of the dead to the Underworld, where they would cross the River Styx with Charon, the boatman, towards Hades' realm) were also worshipped.

In Greek mythology, the dead were allowed to cross into the Underworld only if they were properly buried, and had a coin under their tongues to pay Charon for the crossing (still, considering the number who've died, and even assuming one drachma per person per trip — keeping in mind that those destined to the Isles of the Blessed must take three trips — that's quite a bit of money, and you kind of wonder what Charon spends it on). As for the Romans, the dead joined a group of semi-deities (so that ancestor worship was common) known as Di Manes; thus, it seems that Pluto's realm was not such a bad place.

For the ruling elites, however, as in Egypt, they got themselves godhood! The first was conveyed posthumously to Emperor Julius Caesar, then to placate Caligula (known for his excesses and cruelties) and to Commodus, the son of the brilliant philosopher-king Emperor Marcus Aurelius — a son who was rumored to have killed him (as depicted in the movie *Gladiator*). Somehow, I don't think those godhoods did any of their recipients the least bit of good in the Underworld or on Mount Olympus.

Additionally, when the Romans prayed to their gods, it was not like the "soul-based" religions such as Christianity or Islam, where people prayed to enter Heaven, but people prayed (as it's not uncommon for members of all religions) for mundane rewards, such as prosperity, love, success, happiness; what we all wish for, translated from wishes into prayers.

The idea of the Underworld was nebulous and doubtful to most; the Roman school of Epicurean philosophy taught that there was no life after death, and that the gods didn't interfere in human affairs but led a tranquil and peaceful life that we should emulate — ridding ourselves of fear, anxiety and desire. In many ways, this philosophy resembles the words of the world's classic religious text of the Bhagavad Gita ("The Song by God") as well as those of Buddhism.

The "Gita" is about the closest thing there is to a Hindu "Bible" (though there are many "holy" books). It describes every living being as the temporary joining of atoms; they come together, they come apart. The soul is smaller than an atom; but, if the atheists are correct — says the Avatar of Vishnu, the charioteer of a prince, whose responses define reality — even if there is no immortal soul, still there is nothing to fear: you start as nothing and end as nothing.

According to cosmologists, we were all nothing from the beginning of the universe until a cosmic second ago, and I wasn't the least bit bored in those 13.7 billion years, were you? Yes, there's a difference between not being born and dying — but is there? I remember going for a medical procedure and was given propofol (a common sedative); I asked when the procedure would begin, and I was told it was over. I didn't have the slightest feeling that time had passed, and if I had never woken up…? If our E5 (more about that later) allows you to live a million years, that won't diminish the eternity you'll spend dead by one iota.

Other Romans adopted the Stoic philosophy, which concerned itself with righteous living, with the gods as examples, though it seems to me that the Roman gods wouldn't be good examples for my kids. The most prominent of the Stoics (though it is not known if he was a "card-carrying" Stoic) was the Roman Emperor Marcus Aurelius, whose *Meditations* (never meant for public eyes) are still widely read and quoted to this day.

To the Romans, so far as death was concerned, the soul was immortal, but as in the Hindu religion, it transmigrated to other bodies after death. I guess then it was not strictly "your" soul, like the "soul" of Plato — but an abstract object that enters the body to give life, or intellect — and departs to give it to others. I suppose it is a comfort to know some part of you is immortal — even if it's not YOU.

Later Christianity

The Church became the center of life in Europe at about 500 ACE and its reign lasted to 1500 ACE, when Enlightenment forever changed Western civilization. Petrarch (scholar and poet of early Renaissance Italy) coined the term "Dark Ages" at a time when the prior Roman and Greek civilizations were deemed the "Dark Ages" due to their lack of Christianity. Petrarch judged the 900 years that passed since the end of the Roman Empire the "Dark Ages" because of Europe's loss of the ancients' attainments in the arts and sciences. It was Petrarch and fellow seekers of hidden knowledge that brought about the Renaissance, and eventually, the Enlightenment.

So, what was this Christian world that came to dominate Europe (that is our major interest, as the Enlightenment, which I will discuss further, began there)? For many centuries, it was ruled by the worldview of the Church, first with Platonic philosophy and then with Aristotelian philosophy. Common to both was the immortal soul, a useful device conjured up by the Greek philosophers to replace the logically impossible idea of "resurrection" with the more believable and intuitive idea of an immortal soul. And not only an immortal soul, but one that immediately entered Heaven, Hell, Limbo (for babies and the unbaptized) or Purgatory (for those who were "correctable").

According to Dante (his *Purgatory* never reaching the popularity of his *Inferno* — both from his work *Divine Comedy*), there were pretty unexciting tasks to do in Purgatory, like running at full speed for hundreds of years; it was pretty dull. In *Divine Comedy*, Dante (who was also the main character) had as an ultimate guide the spirit of a woman who rejected him on Earth, finally taking him to the highest Heavens where she dwelt. Interestingly, sex seems to have been Dante's motivation — in his case, love for someone whose characteristics he imagined (though the real woman who inspired him was married and had children, having normal human traits).

What really matters is that for Christians of the Dark ages or later Middle ages — although they were not a static culture with no development — the underlying assumption was that life on Earth was of no real importance, except as a test for the afterlife and your place in it. That was what life was all about. Life was a test that few felt they could pass and many would lie awake at night terrified of the Hell to come.

As the preachers still exhort their flocks, this present life is but the blink of an eye in relation to the eternity that awaits, a "passing shadow", and that "eternity", in bliss or torment, is said to await us all. For me, this present life is the only life we know for sure, and we can expand it to its maximum extent. Does the miracle have to come from a giant old man in the sky, or could it be sent via scientists who have finally solved the riddles of aging?

As we will see, it will be the later one; the hopeless vision of life and death as accepted by most scientists is wrong.

To Nature, the life of the individual is not important, only the life of the species, with individuals being born to replace those dying; like a tree that sheds its leaves to preserve its life, the individual dies that the species may prosper. But unlike in the case of leaves falling from trees, we'll find the aging process is reversible.

The ultimate Biblical dream of humanity immortal on Earth and in the Heavens makes sense when we realize that the "Heavens" are just the universe of which we are part. And the newfound ability to rejuvenate entire animals (including, I believe, humans) gives us the potential of the very long lives needed for a galactic civilization beyond the imagination of the ancients. When the Bible tells us that God said, "My thoughts are not like your thoughts and what I do is beyond your imagining", we must realize that the same could be said by a modern computer programmer to an ancient prophet.

Other Religions

Apart from Judaism, Christianity and Islam (the three Abrahamic faiths), the Hindu faith also ultimately believes in a Supreme God (Krishna, according to the Bhagavad Gita, or Vishnu, of which Krishna is the eighth avatar).

In Hinduism, the individual soul transmigrates according to its Karma (its intents and actions, and their consequences), rising to the levels of gods or sinking to the level of insects, as death forces its transition to a new form in its evolution to rejoin the universal mind, "Atman". This "soul" has no memory of its prior lives; it yet somehow seems to evolve, to learn. The soul is immortal and transmigrates as it evolves to higher and higher forms, until it reaches the highest forms and leaves the "circle of life, death and rebirth". A joining with the God-head (Atman) is the ultimate release; "moksha" to the Hindus, "nirvana" and "satori" to the Buddhist. In this sort of "immortality", the soul is neither body nor mind, neither is it a continuation of personal consciousness, but promises something higher, a joining with the universal mind to become one with all.

Humanity has always sought to join something/someone greater than themselves, whether angels or aliens. So, the search for physical immortality in such a culture of real believers would simply be a distraction, an impediment to spiritual evolution. To true believers in the Abrahamic traditions, a good person should (but not for Homer's reasons) be happy with as short a life as possible, as further stay in this harsh and sinful world merely delays and endangers their "reward".

In truth, the appeal of the later medieval Church was a consequence of the fear of eternal damnation, of a Hell of hideous torment (as Breughel, Bosch and Dante would paint or describe), so that eternal life would be the opposite of comforting, and believers would spend nights awake punishing themselves, all in fear of Hell. So convinced are the faithful of the realities of the afterlife, that many are willing to give up their own lives (and those of others) to claim their reward (and often for their family — there seems to be a myth that you can get better housing in Heaven if you're a "martyr").

Homer's dilemma that life was short, and death (as conceived by the Greek's dull Underworld) infinitely long and tedious, was solved by making death a transition to a better world to come. This still left two problems: Hell and an eternity of suffering if you sinned badly (keeping most people, who are at least occasional sinners, in constant dread), and "faith", the need for a "suspension of disbelief", such that one had to believe in an abstract yet elaborate afterlife with no proof other than the words of "authorities".

I don't believe, though, that any of those "authorities" visited Heaven, Purgatory or Hell. The poet Dante did, but only in his imagination, and only to make political statements about friends and enemies and demonstrate his undying devotion to a woman who spurned him (Beatrice Portinari) and married another. The Greeks, remember, had a special place in Hades for those who wasted their lives pursuing unrequited love — the Mourning Fields. In Dante's case, unrequited love is a motivation for self-improvement, to be worthy of that love. Of course, it seems to me that Dante's love of Beatrice (who guides Dante through Heaven) is a substitute for love of God. It is kind of hard to love an abstract (and perhaps non-existent) entity, except perhaps through His (Jesus, calls God "Father" and not "Mother") creation, of which Beatrice was His most splendid and exemplary product, at least in Dante's eyes.

The "Dark Ages"

For me, there have been several "Dark Ages" — such as the fall of the Mycenean civilization to the Sea People, the fall of the Greek civilization to the Romans and the fall of the Fatimid empire in Spain, which was comprised of Jews, Christians and Muslims working together to understand the world, and faced the invasion of the ignorant and bigoted Moors from northwest Africa who came to determine the fate of that civilization. This fanatically fundamentalist and ignorant Muslim sect separated Muslim and non-Muslim peoples and destroyed their harmony, replacing tolerance with extreme prejudice and resulting in the death of that emerging center of human creativity and understanding.

Another example is the fall of the wildly creative Weimar Republic, in Germany. Also, Alexandria (Egypt) was once one of the world's great cities, where Jews, Christians and Pagans lived harmoniously for centuries, creating a rich and diverse culture — again until religion was used to separate people and another Dark Age ensured.

Even now, we see zealots who see the destruction of hard-won knowledge as their cause, who reject science because it interferes with their beliefs. In almost all cases, these "Dark Ages" were ages where authority and orthodoxy forbid freedom of thought as much as they could by limiting freedom of expression. Expressing views other than those of the authorities was heresy and punishable by torture and death.

So, for example, Nicolaus Copernicus' magnum opus (*De Revolutionibus* — "On Revolutions"), a book which went against Church teachings (though he was himself a high Church official) by assuming the Sun and not the Earth as the center of the "universe", would not be published until after his death and then with the disclaimer in the preface that his Sun-based system was only a mathematical model used to ease calculations (however, the calculations achieved using the heliocentric system are actually much worse than the 1500-year-old *Almagest* of Ptolemy, the mistake being to assume planetary orbits to be perfect circles, which was widely believed, rather than the ellipses they are). However, Galileo was aware of Copernicus' writings and later proved them by showing with his perfected telescope that Venus had phases, like the moon, and thereby circled the Sun.

The Christians started in small numbers in Rome and were regarded with suspicion. Some of their rituals smacked of cannibalism, some of incest to the Romans, so when the Great Fire destroyed large portions of Rome in 64 ACE, it took Emperor Nero no time to accuse, arrest, torture and kill Christians, sometimes burning them as living torches to light his parties, other times having them ripped to pieces by dogs. But Rome had many slaves, many poor and many soldiers, and the religion spread among the people of Rome.

By 313 ACE, Emperor Constantine legitimatized Christianity (after having a vision of the cross before winning a battle against a rival — *In hoc signo vinces*, "in this sign we conquer") by issuing the Edict of Milan, giving complete religious freedom to Romans (Constantine is said to have been converted to Christianity on his deathbed although he always remained the Pontifex Maximus, the leader of the Pagan Roman religion), and seven decades later, Theodosius I made Christianity the official religion of the Roman Empire.

But rather than the religion controlling the state, the state took over the religion, so that the Father, the Son and the Holy Spirit were by pronouncement co-equals (a decision made by a triumvirate of rulers, Theodosius I, Gratian, and Valentinian II, also supposed to be co-equals). Those "foolish madmen" (the triumvirate stated) that did not accept these principles — as many Christians did not — were to be punished by the Emperor as he desired. In 385 ACE, the execution of Priscillian, a Bishop with an ascetic form of Christianity, considered heretical, took place.

At this point, Christianity was transformed from a religion based on the revelations of Jesus and the devotion of his disciples to an official organ of the state. When the ruling elites decide religious doctrines, those doctrines will tend to preserve their privileges (my guess). And though Rome was in decline, it was not helped by waves of immigration by "barbarians", the Goths and the Alan (a Northern Iranian people, their name the dialectal form of "Aryan" [noble], who were a tall, blond people), who were chased across the Rhine by the rampaging Huns — a central Asian people who conquered much of the world, but have left little but the name of their ruler, Attila. Only three words remain of the Hun language, and of the three, one of them might be Alanic (the Huns were another Iranian people).

In any case, I will not consider the influence of Asian philosophies, like Buddhism (Hindus call Buddhists "atheists"), which, except for "mystery cults", like Hinayana Buddhism, makes no mention of life-after-death and is concerned with the devotee's earthly life, without a guarantee of life-after-death.

While in India the afterlife was richly imagined, with millions of gods and supernatural beings, in general the Chinese did not believe in an afterlife or the persistence of life after death. Their emphasis, like ours, was the prolongation of life. Ginseng and astragalus are the two most famous "life-extending" drugs from the Chinese pharmacopeia, but no longevity records have been certifiably reported by the Chinese. Chinese medicine is based on a balance of the opposing forces of yin and yang to promote health. This "theory" of aging (or illness) in practice had no significant effect on lifespan.

Religion is a way of trying to understand and influence the world. A priestly caste developed in all societies, whether shamans, in primitive societies, or priests in medieval society, or scientists in today's society, because it appears that trying to understand and control nature through the use of human intelligence is a feature of all modern peoples. However, any human institution is composed of and by humans, and so it is subject to the strengths and weaknesses of human character — the personal desire for power, wealth, influence and legacy, all have significant effects on those institutions and they can stagnate and rot in place if selfish human authorities, rather than evidence, reason and an ability to learn from experience, decides right and wrong. Once authorities consider as fact that which is not fact, then the forward progress is blocked until such untruths are unlearned, and this will be fought by the powerful elites who first decided those "facts".

The "Middle Ages" were a time during which the Church (Holy Roman Catholic Church) ruled Europe; through the power of excommunication, the Pope, the "king" of the Church (as Cardinals are called "princes of the Church"), could bring kings crawling on their knees. While the "Middle Ages" were supposed to last from the 5th to the 15th centuries, the "Dark Ages" were usually said to last only until the 10th century, and later events occurred which eventually led to the Enlightenment, and a return to seeking knowledge beyond the Bible and the teachings of the Church Fathers.

However, humanity did not change with Enlightenment, nor was everyone enlightened (even today, as can clearly be seen in news headlines). Even the post-Enlightenment enterprise of science is composed of men and women with the same human foibles that affect the new enterprise (now business?) of science, as elites, and not reason and evidence, shaped modern science.

A fairly well understood and predictive science like physics can rely on many "theories", because they are scientific theories. That is different from what we call theories in common language. A theory is an overarching explanation of many confirmed hypotheses. That is, the theory of gravity is a theory because it explains the path of a snowball or the orbit of a planet or why plants grow upwards and downwards. The present theories of aging are not theories in that sense as they make no predictions, or few and of little significance to someone like myself who is searching for immortality, and not just extended old age. I believe that I have found the solution, or at least the first effective solution to aging and I will show you some exciting evidence later — but before we get to that, let's talk a little about the physical basis of life — as that ultimately determines the death of an organism.

2

What is "life"?

> "What drives life is thus a little electric current, set up by the sunshine."
>
> **Albert Szent-Gyorgyi**
> (Hungarian biochemist and Nobel Prize laureate)

When I ask "What is 'life'?", I'm not really being "philosophical" — I'm not questioning the meaning of life; well, it has a meaning to me, to Nature perhaps another meaning, to you? As a biologist, I know that one purpose for every member of a species is to reproduce or to assist in reproduction to keep up its species numbers (not necessarily each member reproduces; in many animal species, like ant and bees, breeders are a tiny fraction of the population). Survival, not of the individual but of the species, is the measure of success in the biological world; many species have been called into existence, but few remain.

Thus, for any and all species, reproduction is a fundamental purpose because there is no life without it. So, one purpose of life is to perpetuate itself. The history of life on Earth tells us that another purpose is for life to occupy all available niches, and in doing so, create their own niches (habitats and habits).

Life is a process found only in living things, but how would you describe that process? Reproduction is sometimes part of it, but it may occur with or without it — so what is happening? The simple answer is that a living organism takes matter and energy out of the environment to maintain its activities and components and make more of itself. In order to accomplish those goals, sources of energy and matter are needed. Sources of matter are obvious for animals: food.

Everything is made of matter, in the form of atoms, themselves composed of *protons* (positively charged particles) and same mass but neutral *neutrons* (both composing the nucleus of the atom), surrounded by a number of *electrons* (negatively charged particles) equal to the number of protons. The number of protons is the "atomic number" of an atom — it defines what type of atom it is, an atom of carbon or iron or oxygen. The atom has enough neutrons to stabilize the nucleus, as all protons are positively charged (so they naturally repel each other — the nucleus would be unstable if forces stronger than the electromagnetic force weren't holding it together).

Still, there are many unstable atomic isotopes (atoms with the same atomic number but different numbers of neutrons), and such nuclei may break apart in unpredictably different ways (except statistically), and at unpredictable times (again, only predictable as a probability). So, if an isotope has a half-life of 12 years (like tritium), it will statistically decay (decompose) by half in 12 years. Tritium is a heavy isotope of hydrogen; it is called "tritium" because it has three particles in its nucleus (two neutrons and one proton) with atomic number 1 (meaning it's a hydrogen atom) and atomic mass 3 (meaning it has two neutrons in addition to one proton). As in all neutral hydrogen atoms, there is one electron circling that nucleus; for most purposes, we only need very simple models of atoms and molecules (joined groups of atoms).

An atom that has the same number of positive charges (protons) as negative charges (orbiting electrons) is, from a distance, electrically neutral. In spite of the fact that the electron has one-thousandth of the mass of a proton, the electrical charge it carries is exactly equal and opposite to the charge on the proton. But electrically neutral atoms are not the only combinations of neutrons, protons and electrons that exist for many elements. Hydrogen atoms, which (mostly) consist of one proton and one electron, can lose their electrons, if they are grabbed by atoms or molecules with a greater affinity for them, or simply by being heated hot enough to knock electrons out of their orbits.

In these cases, we have an "atom" with a number of protons different from the number of electrons. When that happens, this atom with charges (either positive, missing electrons, or negative, with more electrons than protons) is called an ion. The naked proton that remains if the electron is stripped away from hydrogen is called a hydrogen ion (H^+). For a more improbable and energetic form, when the single proton has two electrons (i.e., two negative charges) orbiting about it, it is the negatively charged ion called the "hydride" ion (H^-). Both ions are important actors in our drama.

Don't believe, however, that only hydrogen atoms can lose or gain electrons — sodium, potassium, magnesium, calcium and chloride ions play vital roles in life's machinery. And not only single atoms, but charged groups of atoms like sulfate (SO_4^{2-}) and acetate ($C_2H_3OO^-$) play vital roles in living processes. Even more energetic substances (higher potential energy, i.e., more unstable) are formed by processes we will discuss later where single electrons are "inadvertently" passed to oxygen or nitrogen atoms forming "free radicals", which are extremely unstable and will pass their electron to any molecule nearby.

The other important component is *energy*. So, this is harder to explain because it is abstract, and since "atomic energy" has nothing to do with living processes, we'll just make two easy "classical" physics definitions:

1. Energy is the capacity to do work.
2. Work is equal to the force applied times the distance through which it is applied.

Not so easy? Well, the first part of the definition seems easy enough — we all know what work is; it's what you'd rather not be doing but you do it for the pay. And you know it takes energy because you're tired at the end of it. But here we define energy in a different, simpler and more measurable way (because everything depends on measurement in science).

So, let's say you have a ten-pound (weight) book lying on the floor and you want to place it on a narrow shelf five feet above the ground. Lifting that book off the floor actually requires a bit more than ten pounds of force, because you are accelerating the book from a velocity of zero, when it's on the floor, to some velocity more than zero as it's moving up. According to Newton's laws of motion (they're 400 years old!), if you move the book up at a steady pace (no acceleration) then there should be no net force. After the initial acceleration, the ten pounds of force you exert upwards is exactly equal to the ten pounds of force (its weight) the book exerts downwards. So, you exerted about ten pounds of force over a distance of five feet — and according to statement two, ten pounds (the force you applied) times five feet (the distance through which you applied that force) equal 50 foot-pounds of energy. And that's right.

Now we place that book on the edge of the shelf so that if you placed a feather on the outer border of the book it would fall down (none of which has anything to do with the energy we used, 50 foot-pounds, to get the book up there). So now that energy expended in getting the book up on the shelf is gone! Everything for naught — but not naught! Because the very first law of energy (in the pre-Einsteinian universe, but good enough for most biology and chemistry) is this: energy can neither be created nor destroyed. The way physicists refer to the fact that you cannot create energy and you cannot destroy it either is that energy is "conserved".

So, now we say that the same 50 foot-pounds of work (energy) we expended lifting the book are still stored in that book (according to relativity, its mass is greater [immeasurably so] too). In this case the energy isn't apparent — it's "potential energy".

Now, my task is to break open a walnut — with a feather. Though I try dropping the feather on the walnut a hundred times, it's a tough nut to crack. But I have an idea. I place the walnut where I figure my ten-pound book will land when it falls off the shelf, and then I drop that same feather on the book. Immediately the book tilts and falls, faster and faster until it hits the nut and cracks it. If you could measure all the energy used in breaking the nut, producing the loud clap when it does so, shaking the floors (and ever so slightly the walls and ceiling) and add it all together — I think you already know it will add up to 50 foot-pounds.

In this whole process, two kinds of energy were used. The first kind of energy was applied through moving a massive object over a distance, so motion is implicit in it — zero distance, zero energy — and we'll call it the "energy of motion", or more normally, kinetic energy. The other, potential form of energy, is called by most physicists, not surprisingly, "potential energy".

When you lift something up, you are storing gravitational potential energy, because you are doing work against the gravitational "field"; if you try pushing two objects with the same electrical charge (like charges repel each other, with a force that grows stronger as they get closer) then that too stores potential energy, because if you release those two charged objects you've forced together, they'll fly apart, turning their electrical potential energy into kinetic energy.

So, now, imagine you had a balloon that would not allow charged particles through its walls, and you filled it with hydrogen ions (H^+). What do you suppose would happen?

Well, since all of those hydrogen ions are positively charged, they would all repel each other and the balloon would swell and swell — it wouldn't burst, we'd ensure that, at least under normal conditions. So really it would be like a blown-up balloon, wouldn't it? If it had a thin neck that you were holding clamped while the balloon swelled, what do you suppose were to happen if you released your grip?

You wouldn't be surprised if that balloon were to zip off like a balloon containing compressed air would. Again you'd be turning the energy you provided to force those hydrogen ions into the bag (each new one was harder than the last as the inside of the bag became more positive and more strongly resisted your putting of a new positive charge into it) into potential energy once the balloon was filled and clamped tight. Once you let it go, the potential energy accumulated became the energy of motion, kinetic energy. If you were to hold the balloon in place when you released your clamp, you could get that stream of hydrogen ions to turn around a pinwheel, couldn't you?

What I just described is in principle the very mechanism a bacterium uses to spin its flagella (whip-like organs used for locomotion), or the higher cells — our cells — use to produce ATP in their mitochondria. The entirety of life runs, as the Nobel laureate Albert Szent-Gyorgyi quoted before said, on a little current of electricity, driven by sunshine (through photosynthesis). For animals, however, and for mammals like us, the flow of electrons is from complex food molecules, where they have high potential energies, to oxygen (to form water), where they have very low potential energy. The difference in potential energy is converted into useful work, and worse-than-useless entropy — and entropy can be reversed by the input of energy.

"What I cannot create, I don't understand."

The meaning of this phrase by Richard Feynman is very clear. If I understand the relations between the length, density and tension on a string, I can build a harp; then perhaps, knowing a bit more of mechanics, change that to a harpsichord or piano. But if you don't understand these things, you could never build a piano. Fortunately, we are sometimes lucky and "build pianos" first and then find out how they work — that's the way it worked with electricity (used well before the electron was discovered), and I believe this to be the case with E5 (the treatment responsible for rejuvenating the rats in our experiment).[1]

So let's perform a thought experiment and try to build a living organism; and to simulate life (but not very closely), we can assume any possibility hinted at by current technology as possible, or possibly possible. For example, the use of robots to shape metal into useful parts and assemble those parts into a functional structure is already present in automobile assembly plants the world over. So, let's take that as our goal, to create a mobile automated factory that manufactures mobile automated factories. As for an energy source, let's assume solar energy is used because we're more familiar with the sun at Earth's surface as an energy source supporting all surface life, providing the electron current (electricity) that life runs on. In essence, we actually run on batteries, or better, fuel cells like our machines — the flow of electrons of high potential energy to lower potential energy is the source of life's energy (in plants, sunshine provides high energy electrons, as photons from the sun cause chlorophyll to provide them). So our factory-making factory (which makes factories that make factories), could certainly be solar powered.

Since we don't know how life started, we will start with our factory-making factory pre-assembled and with all the equipment and machinery needed to make every part of itself — including all instructions for how to do so. Now, the first thing we need is energy, because nothing moves without it. As already mentioned, we will use solar energy (solar technology actually provides a higher conversion of light to electrical energy than plants do). However, as the sun doesn't shine all the time, we need to store potential energy for those times the sun doesn't shine (about half the time). So, we will need a storage system built for the predictable, daily "circadian rhythms" with a backup for long-term shortages (in case of storms and other energy-disrupting events).

During the day, when sunlight energy is supplying the needs of our factory, the factory moves and uses its sensors to detect the various minerals it needs to make more of itself, taking in those materials and energy from the environment, and maintaining itself. It stores enough energy that when night comes, it's got enough energy to extract, shape, and assemble the parts needed to build a duplicate of itself. If some parts get worn, it has the ability to make perfect duplicates of any part. And, due to the forces of entropy, it produces during the day lots of heat, in getting and using energy, and this heat (entropy) must be gotten rid of by a built-in air conditioning system, supplied by the stored energy. By the time the sun comes up again, the factory has made and assembled another factory with the energy and materials it accumulated during the day, and returned itself to the same state as it had in the morning.

This proposed process appears to be a very early process with the Bacteria and Archaea domains, which dominated life from about 3.5 billion years ago to about 2.7 billion years ago, when the eukaryotes arose.

However, let's stop at some point about 800 million years before, at the first appearance of a living organism recognizable as such to us to see life as it was. Our model of life, the factory-making factory, is not very different than the real unicellular bacteria and archaea that WERE life for nearly a billion years. Our model does not make the mistake that many biologists of aging make, that there is no need for aging, death and the ultimate capitulation of life to entropy and loss. In fact, since life first appeared on the planet, that first living organism being tiny and perishable and of a complexity beyond anything else on the planet, it defied the biological interpretation of the law of entropy, not by becoming rarer and simpler, but more complex and more common. The complex process of life now involves the conversion of matter to forms that developed self-awareness (and the awareness of death). Living things have grown and diverged to occupy every realm where energy and matter are sufficient to support them. If this interpretation of entropy were applicable to life, nothing would be now alive, as even mountains have been worn away into plains, but life went in the other direction towards greater complexity — I believe that the development of complexity is a natural law of evolving organisms, because for every organism that comes to form a new niche for itself, it forms a new niche for others.

So, what are we missing in our factory-making factory that distinguishes it from the earliest living organisms? We have an energy source, and storage capacity. We make everything needed based on instructions and feedback from the environment carried by sensors. We have a basic system that, being an open system — one that exchanges energy and matter with the environment — is not under the jurisdiction of the law of entropy, which applies to closed systems (that do not exchange matter and energy with the environment). What might be the reason that such a system would die? Lack of energy (say a nuclear winter), or lack (depletion) of needed minerals within the ability of our factory-making factory to harvest. However, other factory-making factories will continue to find and extract minerals and produce more factories. What could possibly cause aging if defective parts are replaced, or "death" if material and energy continue to be provided? The answer, at least in ancient organisms, like gram-positive bacteria (a "simple" structure which very much resembles factory-making factories as their sole purpose is to create more of themselves), is nothing; they are immortal.

Our artificial lifeform bears all the attributes of life — its information, stored on thumb drives or DNA, is replicated (with much more fidelity in electronic systems than actual DNA replication), but is potentially subject to both inadvertent change and deliberate change (altering the program for specific processes or environments), and so is "mutable". It obtains its energy and matter from its environment. Individual factories might cease production (die), but many will still be working, replicating themselves in regions with the resources to support them. However, apart from bad luck, there is no reason for aging (which we define as loss or damage) and death.

As we will see, this artificial life we've created very much resembles in its "life-history" that of the simply prokaryotic domains, Bacteria and Archaea, where death by aging is unknown. So how, why and where did death enter the picture? We've seen that the suffering of damage as a natural entropic process can be overcome in an open system, where both energy and mass can enter or leave the system. As in an air conditioner, which transfers heat from cold places to warm ones, the input of energy works against entropy which would allow heat from the hot outdoors to eventually penetrate the cool room and make it hot — as would be "natural" (an insulator doesn't stop heat from entering a room, merely slows it down).

We saw in our factory-making factory that the heat created by the functioning of the factory had to be gotten rid of or our factory would eventually cease performing; however, in an open system where energy and matter enter, it's not a problem to repair what is broken or actually decrease entropy as illustrated by the air-conditioner, contrary to the proposals of the "old school" of biology that entropy is the reason for death — as Leonard Hayflick, the very person who showed that cells have a limited number of divisions after which they die or become senescent, "explains" in his essay *Entropy Explains Aging, Genetic Determinism Explains Longevity, and Undefined Terminology Explains Misunderstanding Both*,[6] a statement of incredible arrogance, based on misunderstanding basic physical concepts, mainly that aging is entropy and that the diseases of aging have nothing to do with aging.

We have shown that several conditions that result in the diseases of aging could be reversed, for example with a large increase in grip strength only days following E5 injection, or the decrease of inflammatory factors to youthful levels in the same time (more on that later). As inflammatory factors are a presumed cause of many of the diseases of aging, those diseases dependent on them should be prevented.

So, are our "factory-making factories" the same as living organisms? If not, why not?

First some words of praise for the real-life equivalent of the "robots" that operate in our factory-making factory which sort of acts like a 3D printer as it can make any part you need or want — but it's much cleverer than most 3D printers as it can change the properties of the material it excretes to form a three-dimensional object, or a four-dimensional object, if that object does something on its own, like an enzyme breaking apart or putting together specific molecules as the enzyme changes to a different state (which usually means a different shape [conformation]) in working with the dimension of time to accomplish its tasks.

This universal part-generating robot exists in all living things; it is called a ribosome! This machine (it is not alive) is fed instructions from the master set of recipes called DNA via a throwaway copy of its complementary RNA which is sent to the ribosomes in the form of what is called "messenger RNA" (mRNA). Then, the ribosome strings together amino acids (20 different kinds), with different properties (some are positively charged and others negatively charged, some prefer oil to water, others water to oil, some are aromatic [don't worry if you don't know what that means, but "smell" is not their important feature], some aliphatic), into proteins. Some proteins are structural, like the beams and girders, nuts and bolts of our "factory-making factories", and others go on to form those "molecular machines", the enzymes mentioned earlier, which are like the specialized robots and sensors of our "factory-making factories".

Surprisingly, the ribosome is made of three or four large RNA molecules that fold themselves into compact three-dimensional shapes. In simple organisms like bacteria and archaea, there are about 50 proteins attached to this RNA framework (in eukaryotes, like us, about 80 proteins, except in our mitochondrial ribosomes which are more like bacterial ribosomes). Also surprisingly, those ribosomal proteins can be removed, so that just the RNA part (on which those proteins are all assembled) catalyzes the addition of amino acids to a growing chain (polypeptide), based on mRNA (messenger RNA), albeit much more slowly than the intact ribosome. This, among other things like ribozymes (enzymes made exclusively out of RNA) and self-splicing introns,* indicate a world based on RNA that may have existed before proteins became a building material of life ("The RNA world").[7] The ribosome bridges the worlds of RNA and proteins.

Now, of course, in the real world of living things, an organism is enclosed in a membrane, both to keep what is necessary inside the organism and what is inimical, outside. And clearly there must be ways to get energy and matter (nutrients) into the cell and the body, and wastes out of the organism and its cells. In the case of cells, that would be through the cell membrane: a multicomponent structure based on a doubled-film of fat molecules with proteins either floating on the outer or inner side of the double membrane, or penetrating through both layers (sometimes those penetrating proteins have tunnels in their centers that allow only the right sort of molecules in or out of the cell).

Sometimes these processes are passive, if there's more of what's needed outside the cell than inside it; in that case, the "diffusion" (the natural process by which molecules move from regions of higher concentration to lower concentration [they "spread out"]) is called "facilitated diffusion" because it is made faster, "facilitated", by proteins that specifically allow those molecules to enter. Sometimes energy must be applied to force molecules against the concentration gradient (i.e., going from regions of low concentration outside the cell to high concentration inside the cell), and this is called "active transport". Special proteins are used for active transport — and those proteins are part of an entire system for processing those imported molecules. All these devices could be simulated in our factory-making factories.

---

* Introns are non-coding regions of an RNA transcript, or the DNA encoding it, that are eliminated by splicing before translation.

Early life

As I stated earlier, we will not consider the question of how life first arose; as per Feynman's suggestion that we don't understand something until we can build it, we have never in all our attempts to do so created a living thing *de novo*. Arguments still exist as to whether metabolism or genetics came first: did the flow of energy create these systems that drew energy from it, or did molecules with memory come first ("the RNA World") and later latched onto and incorporated environmental energy to its own purposes? At least we know for sure that the first living thing was a self-organized transmutation of environmental energy and matter that could make more of itself.

Thus, the first living organisms were very much like our factory-making factories; they were called gram+ (gram-positive) bacteria because of their thick layer of peptidoglycan (part protein, part carb), which protects the bacterium from changes in osmotic pressure (however, there are bacteria called mycoplasmas that don't have these thick cell walls, or any cell walls, and still do okay) but still allows molecules to enter. Lining the inside of that thick cell wall is the double-membrane discussed, with its myriad of highly controlled entrance and exit ports to adjust the flow of molecules into and out of the cell — mostly food in and waste out, but many cells in higher organisms have specialized functions, such as making hormones, or conveying signals across the body. Interestingly, it is mostly specialized proteins, which are cell-type specific, that vary with age.

Many bacteria also secrete "exoenzymes" to digest food outside themselves, allowing only the small molecule products of that digestion to enter as food, which pass through the cell wall and are selectively taken in (sometimes using stored energy to do so) by protein-based molecules that penetrate the plasma membrane (bacteria and archaea, in general, have the same structure, they have a cell membrane that separates the cell from its environment and then the whole is covered by a thick, protective cell wall).

Competition and death

Death entered this world because of increasing populations and limited resources — competition is the biological mechanism that results from it. Imagine then that our factory-making factory was of one type, but perhaps there were other factory-making factories that were bigger or faster than ours. What to do? Perhaps some modification of our exoenzyme* or waste disposal system might result in a product toxic to our competitor, even though we were protected. We have the code and can introduce random changes. There is evidence that when a bacterium is placed in a stressful environment, it becomes more mutable.

Still, the factory doesn't require OUR involvement, it's all automatic — it introduces random changes, until there's one that produces a toxin that kills off the competitor, and suddenly that one factory-making factory is able to out-compete its larger/faster competitor. And to this day, gram-positive bacteria use toxins as a tool to dominate their tiny playing field (which may be us).

It is believed by some that this chemical warfare between gram-positive bacteria may have had the unintended consequence of creating the new domain of life called Archaea, with the much denser cell membrane, in which branched lipids, instead of linear ones, are more stably attached by ether bonds rather than the less stable ester bonds that hold together unbranched lipids of gram+ bacterial membranes (and our own).

But the most interesting thing about these Archaea is not their pseudo-peptidoglycan cell wall or their different cell membranes, but their DNA, which, like all organismic DNA, is divided into genes with controlling regions (places that repressors or activators can attach, to change the rates of transcription of the RNA of a gene), but unlike bacteria, the archaeal genes are often interrupted or split into parts separated by intervening meaningless sequences of nucleotides (which form the sub-units of nucleic acids, DNA and RNA, as they are strung together), so that in order to get the right instructions to make a component part (a protein), you have to remove the intervening, non-coding DNA sequence.

What has not been explained yet (so far as I have seen; the literature is immense) is how that might have arisen, but you can imagine that gene sequences might also be targets for some toxin, or a self-splicing intron (one carrying its own reverse transcriptase — an enzyme that makes a DNA complementary to an RNA).

There is also evidence, though indirect, of a form of oil that could only come from the ether-linked branched hydrocarbons of archaeal cell

---

* An exoenzyme, or extracellular enzyme, is an enzyme that is secreted by a cell and functions outside that cell.

membranes (and sometimes even containing cyclohexane circles) being found in oil shales said to be 3.8 billion years old, which would make archaea life's oldest form.

Cooperation and Complexity

In 2010, sediments from a deep-sea hydrothermal vent named Loki's Castle were found to contain a variety of novel Archaea, including the "Asgard" group (Asgard was the mythic realm of the Norse gods), containing a novel phylum called Lokiarchaeota, which, in terms of DNA sequence, contains typical archaeal genes and bacterial genes (there is a lot of "horizontal gene transfers" between bacteria and archaea, as they grab useful genes from each other and from between themselves).[8] However, not belonging to either group, membrane-associated genes were found, like actin, that are responsible, in eukaryotes, for causing the membrane indentations that allow them to engulf other organisms. Finally, it wasn't until 2020 that this organism could be grown in pure culture (a process that took a year to see turbidity in their methane-supplied cultures), as the initial density was low and the archaea have a doubling time of two to three weeks, rather than 20 minutes like *E. coli* in rich media.[9]

However, this new form could and perhaps did combine with gram-positive bacteria to form a new sort of cell, a combination of gram-positive bacteria forming the cytoplasm and archaea forming the nucleus of a new domain of living things, the Eukaryotes, the domain to which we belong (Yay team!). One member of the Lokiarchaeota group is *Candidatus* Prometheoarchaeum syntrophicum strain MK-D1, which naturally produces hydrogen as a by-product of its metabolism, and is found in a ménage à trois with a sulfate-reducing bacterium (it passes its high energy electrons from hydrogen gas to sulfate), and a methanogen (an archaeon that uses hydrogen to make methane [just as some do in your intestines to make the "gas" we produce there]). The MK-D1 strain also has a complex membrane incorporating eukaryotic proteins.

The point is that here we see an association that doesn't hurt the MK-D1 strain, but provides a livelihood for the two syntrophs (organisms that "eat together"). But these hangers-on are needed by MK-D1 for both energy intake and production of necessary chemicals, and so the whole is greater than its parts.

Interestingly, the candidate archaeon MK-D1 when finally isolated (after seven years of effort) looked nothing like expected: it did not have the cytoplasmic inclusions expected, but various membranous vesicles extruded from the body like so many arms, forming a new theory about the incorporation of bacteria-based mitochondria and chloroplasts into eukaryotic cells by phagocytosis (cell eating [engulfing]). Also, considering the small size of MK-D1 (about half a micron), the idea of simply surrounding the central MK-D1 with membranes to form a nucleus makes for a believable scenario for two symbionts.

Thus, we see in Nature an increasing complexity in living things through cooperation. When two organisms cooperate to form a third, that new organism is both a possible source of prey to some animals, further increasing their complexity, and a possible competitor to others, thereby requiring the competitors to compete. So, it seems inevitable that evolution brings complexity.

Another example of cooperation concerns the cyanobacteria (once called "blue-green algae") which form strings of cells, all held together by a single sheath. Of this string, one of the cells becomes an anaerobe (while the rest are aerobic and use atmospheric oxygen), and the anaerobic cell, which is conspicuously larger than the other cells, called *heterocyst*, "fixes" nitrogen — takes nitrogen from the dissolved gasses in water and converts it to a form usable by living things, an enzymatic process that cannot occur in the presence of oxygen. The aerobic cells of the cyanobacterial "plant" use sunlight for energy, like our factory-making factory.

Here, we have the beginnings of multicellularity — the cells being bound together, some differentiated for specific functions, and therefore, mutual dependency — as heterocysts need the products of respiration, and the aerobic cells supplying those products need the nitrogen fixed by the heterocysts. Hence, in miniature, we have an organism.

These cells communicate through feedback mechanisms to decide which will be the heterocyst (an irreversible process). Some cells may differentiate into tiny, invasive forms called hormogonia with surprising properties that allow them to propagate far from the "organism" that formed them. It is very similar to the reproduction of higher organisms (though no sex is involved), but it is probably a way to escape impending doom as the tine hormogonia cells revert to normal cyanobacterial cells when they reach a suitable environment. We'll see other instances where cells cooperate to death to ensure their descendants survive.

So, is cyanobacterial sheathed filamentary structure simply cells? Since differentiation is required for the provision of nec available nitrogen, the heterocyst acts like an organ; if it is gone, held together by a common sheath will die (unless another beco erocyst fast enough). So, the picture created is of a higher-order structure. A cell (other than the heterocyst) can die — but that's simply a damage as the other cells remain intact. But, if the heterocyst is the cell that dies, then all the cells of the filament die. Thus, the filament is an organism composed of bacterial cells, *as the fate of the organism can differ from the fate of its cells.*

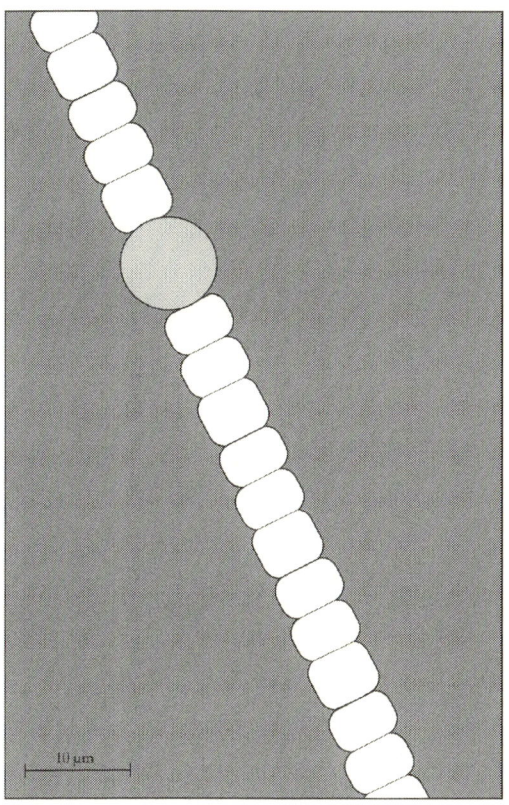

Figure 1: Illustration of a cyanobacteria filamentary structure. The circle in the middle represents the heterocyst, which is attached to the other cells.

The myxobacteria are yet another even better example that is almost the exact equivalent of a eukaryotic "cellular slime mold". Under certain conditions (harsh ones, you can be sure), individual myxobacteria gather to form a stalk and a fruiting body, as some bacteria differentiate into stem-forming

, others into the globular sporangia filled with the spores that still other myxobacteria differentiate into.

Now, patient reader, you might wonder why I'm taking you to such strange places, and the reason is that I want you to see the difference between a cell and an organism, and how the death of a cell can benefit an organism.

But first, we will go back into history, this time more modern history, to see the trail followed by the scientists of the 20th century, which was founded on "theories" based on speculation, authority, false assumptions, and commonsense, which — like the similarly reasonable and commonsense notion that the Sun circles the Earth once every day — were wrong. Though that trail did not lead to the expected prize — biological immortality — it produced some interesting and useful observations on the way. Later, we will go back and decide that those basic assumptions were indeed reasonable, but wrong, and we'll see that following a series of clues based entirely on confirmed and "accepted" evidence resulted in putting us on the correct path, ending in the discovery of a treasure beyond all treasures.

3

Some see "the light"

Were you either a "fool or a madman" to believe anything other than God, Jesus and the Holy Ghost had equal authority? Yes, you were, and by imperial command and by the command of the Holy Roman Catholic Church when the Roman Empire coopted Christianity by about the fifth century (at the time, ruled by the mentioned triumvirate of emperors) as a tool to rule, with both carrot and stick, to both kings and peasants across the Roman world. Even ancient Roman holidays and popular pagan practices were re-packaged in Christian wrappings (including Christmas wrappings), so the ceremony of the winter solstice became Christmas, and the vernal equinox became celebrated as Easter. The Vestal Virgins, an important leadership for the distaff side, became the Holy Sisters and their Mother Superiors, committed to their "sinless" marriage with Jesus.

In fact, the Church became the image and the model for worldly hierarchies (or vice versa), with a Pope, followed by cardinals (princes of the Church), archbishops (the "dukes" of the Church), etc. Another line of the non-commissioned was headed by the Monseigneur, a sort of master-sergeant who kept the troops, the Brothers, in line. Organized to be ruled, the majority of the Church's worshippers believed what they were told, as the sacred texts themselves were unavailable to the laity as they were written in the universal language of the elites, Latin (and Greek and Hebrew as well for the erudite).

The Renaissance

Yersinia pestis is a gram-negative bacterium that is somewhat more complex than the gram-positive bacteria we spoke of. Gram's stain measures the thickness of the peptidoglycan layer. If the layer is thick, the stain remains trapped in it even after washing; if the layer is thin, or non-existent, it doesn't, and such bacteria are *gram negative*. But gram-negative bacteria have something better — they have two cell membranes, and between them, a "periplasmic space" lined with a thin layer of peptidoglycan. The outer cell membrane has protective lipopolysaccharides (which are sometimes lethal to humans, because they cause "toxic shock syndrome"), and pores formed from the protein porin that help the cell preselect what enters into its periplasmic space, and that gives the inner plasma membrane a chance to select again (while defensive enzymes in the periplasmic space destroy unwanted molecules that enter).

However, these bacteria are not in any fundamental way different than our factory-making factories; more bells and whistles perhaps, but they can still at best only make more of themselves. So why am I discussing it here? Because it is the cause of Italy's (and eventually Europe's) Bubonic plague, which I and others believe to be the possible cause of the Renaissance, the Scientific Revolution, and the social revolutions that define modern culture. That a plague can do this may resonate with many readers today.

The intervention of the religious practices was totally ineffective at that time, and monks in particular died in vast numbers as they lived under conditions conducive to the spread of the bacterium *Yersinia pestis* — which was spread by the bite of fleas infected with the bacterium, and the fleas were carried by rats infected by the bacterium. The fleas would leave the dead rat and settle on humans to suck their blood, but something strange would happen; the bacteria would have formed a slimy film which prevents food from traveling to the flea's gut — and when it drinks blood, it cannot swallow it (the entrance to its intestine being blocked), and instead it dislodges the bacterial film and vomits it back into the host's bloodstream.

The yersinia lives in many different hosts and has adaptations for all of them. Its virulence to humans and many of our domestic animals is in part carried by two relatively small mini-chromosomes containing only a few genes, circular DNA molecules, called *plasmids*. The plasmids are what makes it deadly, each one carrying genes to prevent white blood cells from eating and digesting the bacterial cells, and allowing them to do what they do to fleas. In fact, these bacteria can penetrate and then reproduce inside the human immune cells called monocytes that are designed to kill them, which then traitorously help them spread throughout the body.

That's why the lymph nodes become prominent in the neck and groin, as infections in the regions drained by lymph vessels occur and the armies of white blood cells start to reproduce. Lymph nodes are a sort of "barracks" for the immune cells that fight off foreign invaders. These nodes swell as they increase their numbers of white blood cells in order to fight imminent infections. When infected with *Yersinia pestis*, large populations of the lymph node's monocytes are infected and the lymph nodes start to swell as the monocytes themselves become factories for the production of the bacteria they're supposed to fight. As a consequence, the lymph nodes might grow to enormous sizes (as big as apples according to accounts of the time) in the armpit and groin and eventually turn black, and burst, painfully, exuding puss and blood. Death soon follows.

Yet, now, modern science can cure us of this "Scourge of God" with a couple of injections — so, is medicine and science against the will of God? Find out how many sick priests, mullahs and pandits pray for death when they're sick and refuse to visit a doctor. My guess is very few, but all of them make great excuses for their need to visit one (for the sake of their parishioners, no doubt), when all anybody should need is to live a good life, following whatever commandments their Lord(s) gave, such as "welcome the stranger", and die as quickly as possible. It works with suicide-bombers.

Even now (2021), in rural USA, the Enlightenment hasn't arrived, with preachers, ignorant of the knowledge gained in the past 2,000 years, entering psychotic states where they babel incoherently (speaking in tongues, or pretending to do so), and periodically predicting the end of the world as revealed to them by God. Of course, it never happens, but their followers know there was a glitch, a slight miscalculation, and the pastor then predicts some future date, with some Biblically valid reason (I guess God made a mistake or didn't communicate too clearly), but that day never comes either. My guess is that the "faithful" give big when they know the end is coming (you can't take it with you). Folks suddenly get religious after a rousing sermon telling them the world's going to end soon.

Sadly, with humanity on a single world (as the physicist Stephen Hawking pointed out), an asteroid that would destroy the Earth would finally be the end of history, or perhaps a history of humanity will be a doctoral thesis on another world. Maybe we are the only intelligent life-forms in the reachable universe and what was a universe of our perceptions and our minds, of stars and planets and nebulae, becomes dead itself, a tree that fell in the forest unheard, lifeless rocks and incandescent gasses. The answer, of course, is to spread into space — and find or make ways to live there. A very, very extended youth lasting centuries is a requirement for a star-spanning civilization where we'll have "room enough, and time".

The Black Death

Twelve ships entered the harbor of the port city of Messina in Sicily, in October 1347, and when they docked, the townspeople were shocked to see the ships filled with dead sailors and those still alive, but very sick, covered in bleeding boils. These "death ships" were sent away, but it was too late — the plague had already spread to Messina. Soon Marseilles in France became a disease center, with the plague spreading across Europe along the trade routes. First in the bigger cities, then in smaller ones, villages and hamlets — though never as deadly as in the big cities. Cities like Milano lost half their population, and in total 20 million Europeans died of the Bubonic (or "Black") plague, about 1/3 of all Europe at the time. It was a branch of the Great Pestilence that was spreading through the Near and Far East at the same time.

However, people at that time did not think of it as we do now; cause and effect did not exist for the phenomena of plagues or celestial events. Everything that happened was the will of God, so if God was punishing Christianity, then that must be because God was angry at Christians (I can't personally contradict that). So, people must stop sinning. But that wasn't enough; heretics and Jews, non-believers in their midst, were their first thoughts as to the reason for God's anger (there's always a better return when attacking small, rich and vulnerable communities — full disclosure, my parents were Jews).

Jews were killed by the thousands, accused of spreading plague and sometimes confessing — under torture, of course. There was a set of skills and technologies (the "rack" was one, and the "Iron Maiden" was a particularly depraved one, but there were worse) developed by the Church apparently to demonstrate people's inhumanity to humanity. The tortures for women were more horrific and usually dealt with their female parts — an education to (voluntarily) celibate monks. It is not surprising that, given such incentives, "weak-willed" Jews admitted to poisoning the wells to spread the Plague. I would suggest not criticizing them without having taken a few turns on the rack first.

Now that we know the cause (and cure) for this disease, we have to understand that all the accusations and all the confessions were simply lies, because the disease spreads by the bite of an infected rat, or an infected flea — or in its pneumonic, and most deadly form, by simply inhaling air breathed or spoken or coughed out by a sufferer. Thousands of innocent people died for nothing. Others, often upper-class people, banded together to expiate their sins, and perhaps those of their audience, when they marched into a town and whipped themselves and each other with heavy leather belts studded with sharp metal. They would do this three times a day, for 33 ½ days and then leave for another town.

However, though the performance of these "flagellants" ("whippers") made people feel better, it did not stop the plague. Actually, the Church worried about the increasing influence of these self-flagellating "irregulars", and forbade the practice, though in truth, the Church itself was particularly hard hit, especially the monks, who (as already mentioned) lived under conditions well-suited for the spread of the disease. In order to keep up the numbers of monks needed to attend to their "flocks", the Church was forced to choose people as monks and nuns that were unsuitable to the profession, and as Boccaccio's Decameron and Chaucer's Canterbury Tales describe, the ribaldry and corruption of the clergy was widely known.

In the beginning, the strategy was an extreme adherence to religious doctrine, and extremist answers based on the belief that the teachings of the Church would save people, but after a while it became clear that the Church had no power to deal with the plague and that its leaders were no more Holy than thou (whoever). Suddenly chinks developed in the Church's overall picture of the world, so that a little light was able to seep in.

The huge losses in population opened opportunities for the common folk, as the nobles, rich in land, suffered as land became cheap and wages high, so that serfs could buy their freedom, and buy their land. Peasants became merchants and merchants became nobles. In fact, the opposite happened in Southern Italy, where the nobles exerted even greater control on the serfs remaining. However, overall, throughout most of Europe, serfdom disappeared, and industries developed — machines were invented to replace the lost human labor.

The arts prospered as the newly rich, quite aware of their lack of "bloodlines" (rich industrialist families such as the Medici and the Sforza), would prove their contributions to higher culture by sponsoring some of the greatest artists in history. And the graphic arts changed as well, as the ascetic symbolic and stilted "Christian" art was replaced by realistic scenes and statues often depicting curvaceous female nudes representing scenes imagined from Greek and Roman mythologies (and quite a lot of flesh).

Well before the Renaissance started in Italy in the 15th century, there were other Renaissances; one, the direct result of King Charlemagne, especially in the areas of learning and education, in the 9th century; still another followed the enlightened kingship of Otto, king of the Saxons and Holy Roman Emperor, whose ecclesiastical duties brought him in contact with the best and brightest of his kingdom. However, these were "top-down" changes, and hence temporary, depending on who was on top.

More significant perhaps were the Crusades, where European Christians, on orders of the Pope (*Deus volte!* — "God wills it!") attempted to evict Muslims from "the Holy Land", what is present-day Israel (upon arriving in Jerusalem, the European Christians immediately slaughtered the Arab Christians that came to greet them, so it was a racial thing even then; another example of this is that they also murdered European Jews in preparation).

When Sicily was reconquered from the Arabs in the 11th century, and Europe now dealt, both peacefully and not, with the Umayyad Empire in Spain, the exchange of knowledge was one way, as the Muslims, both Arab and Persian, and even Jewish scholars, retained many of the works of Greek science and Roman law that were forgotten or lost in Europe. These cultures even expanded Greek scientific knowledge particularly in mathematics, optics, astronomy and medicine. However, that was not the way Europeans saw it. They were aghast that Spain could allow Muslims and Jews to openly practice their religion. Europeans thought Spain was disgusting, Europe was for Christians!

In 1492, Queen Isabella exiled "her Jews" (as Jews belonged to the regents) to European applause, but that did not suffice for the "saintly" Thomas More or Martin Luther who called Spaniards "faithless Jews and baptized Moors". So, the converted Jews (many of whom were believers in Christianity) and the Moriscos (converted Muslims, many of whom remained Muslim) were also exiled. One prominent Norseman compared them to rats.

So, never doubt the racism behind the purportedly "Christian" militancy. The Crusaders did not want to send their Arab adversaries to Heaven, but to Hell. In the same year as Spain expelled its Jews, Christopher Columbus discovered a "New World", full of riches, and a world that the Bible never mentioned. Indeed, learning and power shifted to Europe, with that discovery.

However, the exchanges with the advanced Islamic civilization and the restoration of many Greek and Roman manuscripts by people like Petrarch started to change European thinking in a profound way. Petrarch (his real name was Francesco Petrarca) was a poet who uncovered many ancient and lost manuscripts in Greek and Latin (in chapels and monasteries). He was one of the first to use the term "Dark Ages" in the way we mean them, when during 1638 he realized he was living in a Dark Age, even when it was common to think of the ages of antiquity, those times before Christianity, as the "Dark Ages".

Like Dante, a contemporary, who had his idealized love in Beatrice, a real woman who married another man, Petrarch also had his "Laura". I'm not sure of what that means, but perhaps having a "real" model of the potential for human attainment was then necessary (even though straying far from the reality of the women themselves — maybe if Dante and Petrarch had respectively married Beatrice and Laura, much of the Renaissance might have never happened). Perhaps the old images of a saint or the Holy Mother no longer sufficed to move people's hearts; by this time they were "old fashioned", at least among some of the most intelligent and knowledgeable men and women of that time.

Nevertheless, Petrarch believed himself a good Catholic and never believed that religious faith and human achievement were mutually exclusive. Though a poet of renown, Petrarch's main contribution was to find and popularize ancient classics (finding Latin texts in old monasteries and translating Greek into Latin to give Greek texts a wider readership), and, as we can see from Renaissance art, the world went wild for them. Latin would become the universal language of the literate, and the intelligentsia couldn't help throwing in some Greek into their writings to show their sophistication.

So, again I'm taking you on a journey through history, but really through the history of ideas, and that is what we are looking for — where the ideas that might lead to biological immortality might come from. We know it cannot come from a world where thought is constrained to the beliefs of elites based on their word alone. The "soul", conceived of as a non-corporeal, immortal part of us that survives death, has absolutely no proof. As we have seen, it does not even have the authority of Jesus, who as a Jew of his time did not believe in an immortal "soul", but in the resurrection of the dead in perfect bodies.

Greek philosophers supposed "wisdom" about immortal souls (though rather impersonal ones that find new bodies when their possessors die — in Plato [Socrates] view, and that of the early Church) was implemented as a more suitable alternative to resurrection that fit the "advanced" thinking of the time. But for people to really believe in the promised immortality, it couldn't be baseless, it should be based on evidence and proof. It turns out that some Christians do not believe in immortal souls, and Martin Luther himself contested it (Christian conditionalism), with the rejected resurrected being simply burned up — the ultimate death — and not tortured eternally (Annihilationism). So, while this is not the Church's opinion, the existence of an immortal soul is not a must for Christianity.

On a personal note: how can you believe someone who tells you that if you do what they say, you'll be rewarded after you're dead, and then takes money from you to ensure that? In medieval Judaism, a *rabbi* ("master") was not allowed to make money from his religious duties, he needed to have a full-time job not associated with being a rabbi — a good system I think.

To me, not only does the immortal soul "solve" the problem of death, but it also supports a huge infrastructure that supports that "immortal soul", providing the rules to follow, the teachers to teach them, the preachers to preach them, the rituals needed and the gifts expected in order that the immortal souls find eternal life in the best of worlds. That was the promise of the Egyptian priests when they started mummifying non-royalty. By the time Alexandria was one of the world's greatest cities, rich Egyptians were taught the "Book of the Dead" and memorized exactly what needed to be said to each of the 40 guardian gods they (actually their preserved bodies, i.e., "mummies", a costly process, but absolutely required for immortality) would be challenged by and needed to pass along the way to paradise. However, do you really think their preserved corpses (with brains and other organs removed) are enjoying their Heaven? A lot of money and talent was wasted on a worthless pursuit.

Figure 2: Hieronymus Bosch's depiction of Hell in his painting *The Garden of Earthly Delights* (Museo del Prado in Madrid, c. 1495–1505).

Yet that is the promise of every mullah, priest and pandit, and that's how they make their living and retain their power. All of them have different beliefs and requirements — if you're a Christian and haven't sworn that Allah is the One God and Mohammed is his Prophet (in Arabic), then no matter how good you were (as a Christian or even a person), you are still going to Hell according to Islamic rules, and the feeling's mutual. So, which one is right? Your immortal soul depends on it — if you have one. Back in medieval times, the intellectual supremacy of the Greek philosophers was taken for granted, with their advanced idea of an immortal soul replacing the primitive Hebrew idea of resurrection, but that supremacy hasn't existed for a long time, as we now know so much more than they did. And as you'll see, that makes a big difference.

A little light

First, came new ways of thinking about the world that was (Classical times), and then the world that is. Works of literature, science and mathematics from Greece and Rome (what survived of them) were brought about by the rediscovery of the classics and became widespread and widely known, at least by high society. In an interesting twist of history, the beautiful logic of Aristotle defeated the odd notion of Aristarchus that the Earth revolves around the Sun, rather than what is obviously true that the Sun revolves around the Earth (you can see that with your own eyes). The theory that Aristarchus promoted also required the Earth to spin about its own North-South axis to produce day and night. If Occam's Razor (the problem-solving principle that the simplest explanation is usually the best one) is applied to the competing sun-centered (heliocentric) theory of Aristarchus and the Earth-centered (geocentric) astronomy of Ptolemy (whose calculations, in his *Almagest*, proved fairly correct for 1500 years), it is very clear that the simpler geocentric theory, not having to propose a rotating Earth that no one could feel rotating, is the clear winner.

Furthermore, Aristotle posited that if it were true that the Earth circled the Sun (at a great distance), then the positions of the nearer stars would shift against the further "fixed" (not moving) stars, as the vantage point in summer would be millions of miles (as they calculated) away from its vantage point in winter, six months away and at the other "end" of its orbit (this is called parallax) — in reality, it is 186 million miles away from its summer position. Since the difference was never seen (the stars shifting their relative positions from summer to winter), Aristotle concluded both from evidence and from logic that Aristarchus was wrong. Of course, Aristarchus was correct. Aristarchus' response to that latter assertion about the stars shifting was that the stars were so far away that you couldn't see the slight displacements a mere 186 million miles make. But it took the telescope and the photograph to show that Aristarchus was completely correct, no matter how "lame" his response seemed to the ancients; new technology revealed him to be completely correct, right up until and including his "lame" excuse.

"Nick Cooper", better known as Nicholas Copernicus, thought about the heliocentric theory of Aristarchus and realized that while the current "science" of astronomy with its "epicycles" and "equants" did a marvelous job of predicting the motions of the planets, the Sun and the Moon (again, Ptolemy's *Almagest*, a book with tables of celestial phenomena, was used for 1500 years), and the strange and complex motions ("wheels within wheels") of mechanical and mathematical models produced accurate results, they DID NOT MAKE SENSE. That is why "angels" were assigned to keep the planets in their orbits. What constrained planets to this incredibly complex motion (with epicycles on epicycles)? The "theory" of the geocentric universe produced accurate predictions, but the heliocentric theory produced a rational picture of all planets orbiting the Sun, with no need for epicycles and equants.

Now one might simply ask about the force that holds all planets in orbit around the Sun, instead of the mysterious forces that constrained planets in circular orbits around imaginary points that might themselves be traveling in circles (more circles, more accuracy!). In the heliocentric model, a central force emanating from the Sun could explain it all. One very large problem with that was that the charts — the new heliocentric-based charts produced by Copernicus — didn't work very well. But science was now closer to the truth, and Copernicus came closer than anyone. However, he never allowed his work to be published during his life (he was a Church official, as mentioned), and in a preface to his posthumous work he states that he never intended (Heaven forbid!) that his heliocentric model represents reality, it was only a mathematical model that simplified calculation.

And so it was that the heliocentric model resurfaced, though it only gave approximate answers; nevertheless, as soon as Kepler discovered the real orbits of planets to be ellipses rather than the perfect circles of Copernicus' belief (it was a good guess — it made sense of a phenomenon, and provided a narrative rather than a simple prediction), and finally with Newton's additions, we could use it to predict the motion of planets billions of miles away, for centuries ahead of time.

I also want to mention the Danish nobleman Tycho Brahe, who had Johannes Kepler as his brilliant assistant; between them (and without the use of a telescope) they plotted the orbit of Mars and discovered that it, like the orbits of other planets, was not a circle but an ellipse, and with that information — that the planets travel in elliptical orbits with one focus being the Sun — the heliocentric calculations worked out perfectly.

These were the main points of this story: rather than relying on authority, Copernicus had the temerity to question the opinions of the Church's greatest authority on the physical world, Aristotle, and was later proved to be correct by observation. And later, Brahe and Kepler actually measured the orbits of the planets rather than relying on the well-documented and universally accepted notions of the authorities. They wanted understandable explanations for the real world they lived in. No longer "it's God's will" was the answer to every question.

At this point, Galileo enters the picture to demonstrate how instrumentation (his perfection of the telescope) — not to mention his genius — brought about yet another blow to the Church (though Galileo considered himself a good Christian). Galileo was a brilliant man who became a very rich man from his drive and ingenuity — he was the Elon Musk of his day and more. When the telescope first came to Florence, Galileo missed his opportunity to view it, but based on description and Galileo's own expertise in optics he soon had a working instrument. The telescope then was not the instrument of an astronomer, but a spy-glass, useful for observing enemy movements from a safe distance, or seeing a laden ship coming into port from far enough away to buy shares in it before others knew it was coming into dock. As seafaring was a risky business and many ships set out never returned, both risk and return on investment were high! The telescope alleviated the risk for those who knew a ship was to berth before others and could buy shares cheaply.

Prior to this, the heavens were thought of as Heaven, the Celestial Sphere with God and his hierarchy of superior beings (angels, archangels, powers, dominions, seraphim, cherubim, virtues and other fantastical creatures — "wheels within wheels"), which were inspired by Biblical stories. Much effort was spent in ordering this celestial hierarchy (will we ever know if they were correct?). I've got to admit, reading about the various properties and appearances of these celestial creatures does make me think about "Ancient Aliens". Cherubim, for example were not like (fatty) winged babies, but monsters with four heads — human, ox, lion and eagle — with a lion's body, ox's hooves and four conjoined wings covered in eyes. And why not? The heavens above — the origin of ancient astronauts — were then Heaven, where God and his host resided and to where good "souls" were sent.

However, when Galileo looked at the heavens with his telescope, he did not find heavenly perfection — all things heavenly being composed of imperishable "quintessence" (the "fifth essence" or "fifth element", the other four being earth, air, water and fire). Instead, Galileo found the Moon was a rocky world, with mountains and valleys — he could even gauge the height of those Moon mountains by the shadows they cast. He discovered spots on the purportedly perfect face of the Sun, and that Jupiter had not one, but four moons circling about it (four of 62 known moons; they're still called the Galilean moons: Europe, Io, Ganymede and Calisto). Furthermore, his depiction of the phases of Venus absolutely confirmed the theories of Copernicus and Kepler, as such phases could not occur that way if the Sun circled the Earth.

The Church could not tolerate this much light. Galileo (who was said to have had the sharpest tongue in Italy) was friends with a Cardinal when young, and once he became a hard-liner Pope, he waited until Galileo was 70 and sick and made him recant his heliocentric theory, with instruments of medieval torture spread out before him. Galileo then "admitted" that the Earth does not move (it is rumored that he muttered "It still moves" after his recantation), and was confined for life to house arrest until he died in 1642, the year Isaac Newton was born. During the remainder of his life under house arrest (as I was during 2020 due to the COVID-19 pandemic), he continued researching and produced the first physics text, one used for centuries.

Now, the coincidence between the year of Galileo's death and Newton's birth is surely just that, and I won't spend much time on this very eccentric English gentleman (though he's worth many volumes), because this is not our goal — we're looking at least to lengthen our lives to avoid death, and perhaps achieve eternal youth, that's why you're reading this. But still at that time, though Classical works were prominent and every educated European could converse in Latin, the Church dominated thought.

Newton was himself a Christian, but quite unorthodox (which is why he refused a professorship, as at the time it had ecclesiastical requirements). He used his great powers of thought to try to extract messages "hidden" in the Bible. He worked on alchemy as well, and achieved nothing in those two realms, but he made major improvements in instrumentation (inventing the reflecting telescope), and defined a tiny set of three laws of motion (okay, the first two came from Galileo), and the law of gravity, which taken together made sense of all of Kepler's laws of planetary motion. Nature could now be understood. Four simple laws defined the motion of stars and planets.

Based on the laws of gravity and motion, all the orbits of all the planets were explained (and later new planets were discovered when slight deviations from predicted orbits were observed). Even the path of a comet and frequency of appearance could be predicted (as it was by Haley) and this harbinger of disaster (*dis* [bad], *aster* [star]) became just another predictable celestial object. Even today, Newton's laws are accurate enough for us to land our "rovers" on Mars.

But just as the mystery behind Ptolemy's epicycles and equants disappeared when celestial motion was properly understood, force-at-a-distance disappeared when gravity was properly understood by Einstein, though for most present, practical purposes, Newton's laws continue to serve us well (however, both Einstein's laws of special and general relativity are required by our fast-circling, satellite-based geo-positioning systems — GPS).

The invention of so great an advance in mathematics as the calculus should have been the tallest feather in Newton's cap — but the invention of the calculus is also claimed by the brilliant polymath Gottfried Leibniz (the French philosopher Diderot said that when comparing himself to Leibniz, he just wanted to throw his books away, crawl into some quite corner and die). Historians now state that Newton discovered calculus first but Leibniz published first. Newton insisted that when Leibniz had visited him in England, he stole Newton's ideas about calculus and Leibniz was never able to shake that accusation and wouldn't have been safe in England (a land that worshipped Newton during his life and believed Leibniz to be a thief). The feud never ended and Leibniz had a long and bitter correspondence with Newton towards the end of his life.

This personal battle was to have serious repercussions for England, as the British were so incensed by the purported theft that they isolated themselves from European mathematics, and Great Britain became a mathematical backwater for a hundred years as mathematicians in France and Germany massively advanced mathematics based on Leibniz's calculus (we still use the Leibniz notation, "dx/dy", rather than Newton's notation [an x with a dot above it, $\dot{x}$], except when we talk about derivatives with respect to time).

There are many lessons there too. Nationalism and the deification of the man Newton (a very unusual man who bragged about his virginity, was very paranoid and reputedly enjoyed hanging counterfeiters himself when he was appointed Exchequer of the Mint), led to a sulky, willful ignorance on England's part that hurt it for a century.

Here we see one very interesting thing about science which is its ever-increasing precision of measurement together with a simplicity of description. The complex calculations of Ptolemy produced accurate numbers, but no understanding of why those numbers should be as they were. Copernicus produced numbers that were not quite right (because he retained the medieval notion that planetary orbits are perfect circles — which was "straightened" out by Kepler and Brahe), yet his heliocentric theory eventually won. Why? Because heliocentric orbits created a simpler, more understandable model of reality, an understandable narrative that eventually produced better, and even more accurate results than Ptolemy's, because real understanding produces real results.

Kepler imagined that a force emanated from the Sun and swept the planets along their orbits. Newton then explained this as an attractive force that all massive bodies exert equally and oppositely on each other. A force proportional to the product (multiplication) of the masses and inversely proportional to the distance between their centers squared ($F_g = Gm_1m_2/r^2$, where G is a constant, the universal gravitational constant, good anywhere in the universe, so far as we know), with the Sun being the most massive of solar system objects. Later we found that Einstein's theory of gravity replaces Newton's and explains the apparent action at a distance by exploring how mass shapes space-time — how space and time, inseparable, determine motion. This theory is quite different from Newton's, but in the case of bodies traveling at speeds considerably less than light-speed, Newton's theory is an excellent approximation to the truth — and one that gave us the ability to navigate interplanetary space.

However, Einstein's theory is not complete either as it doesn't explain the quantum world (but that's beyond our scope for now). While the description of how the universe works on a small scale is not simple to understand (unless you accept the multiverse hypothesis), it can actually be written in a couple of lines (that would take volumes to unpack), so in that sense Einstein continued the tradition started by Copernicus, Galileo and

Newton showing that a few simple laws determine the complex behavior of the physical universe.

Can we ever "know it all"? That very much depends on the capacity of the human brain, and as has been reputed to have been said by the great biochemist J.B.S. Haldane, "Not only is the universe queerer than we imagine, it's queerer than we can imagine". Einstein too offered a road to immortality; as you approach the speed of light, your clocks will slow down, approaching zero as you near the speed of light — and, were you going fast enough, a thousand years on Earth would pass while you drank your morning coffee, but subjectively, you'd age and die as normal. So, that's no real path.

Darwin explains evolution

In 1858, Charles Darwin and Alfred Russel Wallace published their *Theory of the origin of species by natural selection*. The theory had a profound impact, not only on science, but on religion and politics as well. This is one more example of humanity looking at the "real" world rather than that of the entrenched Church doctrines. Jean-Baptist Lamarck had already proposed a complete theory of evolution based on the inheritance of acquired characteristics (proto-giraffes reach for the higher leaves and so their offspring have longer necks), but there was no known mechanism for this and there was no evidence to support this assumption — as the activities of the animal by unknown secretions of cells would have to affect its "germ plasm".*

The resolution of this fundamental question by Darwin was based on fossil evidence, on his own discoveries and on his realization — through writings of the English scholar Thomas Malthus — that the Earth has a limited carrying capacity, reinforcing well-known discoveries by the French naturalist Georges Cuvier (1809) that different species that lived in the past no longer lived. Wallace also had an important contribution, as he recognized that new butterfly species (he was a professional butterfly hunter) are always found next to very similarly related species, possibly being more selectively adapted to their habitat.

Physical science and mathematics prospered during the 16th and 17th centuries, with the discoveries of Copernicus, Kepler and Brahe, Newton

---

* Germ plasm is a biological concept which states that heritable information is transmitted only by germ cells in the gonads (ovaries and testes), not by somatic cells.

and Leibniz, which transformed the educated European's world view. Just as Greek geometers such as Euclid, they took a complex subject and transformed it into a tiny set or rules or *axioms* (automatically accepted as logically certain statements of facts, i.e., two points determine a line) which could be used to understand, in an abstract way, the nature of real-world objects. So, the physicists and astronomers were able to resolve the entire universe, the movements of planets and comets, of galaxies and moons based on four simple laws. The desire to apply the same sort of analysis to other fields of human inquiry was obvious to many.

Charles Darwin did not invent evolution. Evolution, as it is currently used in biology, refers to species changes, and is not a new concept. The Greek atomists had early decided that the world consisted of atoms and void, the atoms combining and recombining to make the structures of this world. Leucippus and Democritus (around 500 BCE) are the best-known exponents of this philosophy. Their philosophy was contemporary of Aristotle, but Atomism was not as popular. Aristotle dominated the world of Greek philosophy, so the atomist views were not accepted. The lesson here is that the popularity of a "theory" (conjecture) has little to do with its truth; like a movie, a theory is popular if it leads to a dramatic (very good or bad) ending.

That Aristotle was a great genius there is no doubt, but by being exalted as the ultimate possessor of truth, his words prevented progress in knowledge. What was true — the Aristotelian rules of logic for example — remains true, but nearly all of his biology was wrong, based as it was on the Platonic idea that every creature was a poor representation of its perfect model. In Aristotelian biology, there was no evolution, every species remained the same forever. In computer terms, every individual was a specific implementation of a class. What was needed was a system able to distinguish what was true from what was not. But that had to wait until the Renaissance times to be incorporate in theory by the Lord Chancellor of England (somewhat equivalent to the Chief Justice of the Supreme Court of the USA, but with more authority) Francis Bacon, whose writings were the inspiration for the scientific method. Indeed, all that is required by a scientific "theory" is that it meets the rules of evidence — as in law, evidence is used to decide truth or falsity.

While it was generally understood that matter could be divided indefinitely, the atomists understood that if that were so — if matter were infinitely divisible — then there would be no room for movement or change, so a space or void had to exist. So, the world was conceived to be composed of uncuttable basic particles called *atoms* (which literally means "uncuttable"), and a space, or void, separating them, and they imagined atoms to be in constant motion, bouncing off of each other.

Although the atomists could use their hypothesis to explain some phenomena such as evaporation, that was not their purpose. And life, in the atomists' opinion, arose from natural processes. The pre-Socratic atomists believed life to have resulted from an infinite chaos of atoms which, following natural laws and undirected processes, produced life, in a very similar way to our modern views. And, while their original writings were lost, their ideas were retained by the Latin poet Titus Lucretius in his five-volume work *On the nature of things* (rediscovered and republished in 1417, and so, they were available during Renaissance times).

The atomists believed that Nature (not a demiurge, as Plato believed, or later, an internal "psyche" — the ultimate cause of all life according to Aristotle — but rather undirected natural laws and chance operating on atoms) produced new species "through an undirected process that sorts out the best adapted forms and eliminates those not suited to their conditions".[10] By this description, the ancient atomists not only accounted for evolution in the modern manner (though leaving out the details), but also anticipated by 1900 years Darwin and Wallace's "survival of the fittest" or "natural selection" as the mechanism behind evolution. There's one big difference, however; the findings of Darwin and Wallace were based on evidence.

As close as the atomists came to modern atomic theory and that things people believed were substances, like heat and dryness, were actually sensations caused by atoms, the explanations, though logical (the Greek philosophers all tended to be logical), were just words with no proof; there was no way to distinguish a true explanation from a false one except logic. Our computers, however, have perfect logic (when I worked as a programmer, I had a sign on my desk saying "The compiler is never wrong."), but every programmer knows the expression: "Garbage in, garbage out". The Greeks believed the ultimate truth could be derived from pure thinking, and Aristotle was the master of that (we still use Aristotelian logic today). But it cannot.

Had the early Christians not destroyed the advances in technology made by the Hellenic civilization in Alexandria, as they considered those advances unimportant, who knows what might have been. But it seems it's always ethnic, religious and racial differences that are used by tyrants to separate people and gain power. We've seen though that when different creatures cooperate, they can do more together than the sum of what they could do separately.

Interestingly, the same argument about whether life could arise by chance given its inherent complexity, or whether it was produced by a Creator or a Designer was also argued about in ancient Greece. Both Plato and Aristotle (who, as mentioned, came to prominence again later in medieval Church history as their secular guide), as you might expect, believed in non-material souls (we can only define them as abstract objects, and without the persistence of memory and personality, though the modern Catholic catechism says it retains those attributes).

Also, Aristotle was indefinite about the immortality of the soul. He believed the "soul" was tripartite; two parts (the ones concerned with emotions and desires) would die with the body, but the part capable of logic was immortal. A modern scientist viewing these conjectures about the properties of the "soul" would realize that Aristotle was talking about the human brain and endocrine systems. And yes, thoughts, and therefore the mind, are immaterial, but they have a material basis in neurons and glia in the brain. Shave off the top 2 millimeters of the cortex and see how much logic-capability is left (none).

Yet, those views of the Greeks were accepted by the Church, but those of Jesus, rejected. The views of the atomists, like those of Aristarchus before, were rejected, but on the basis of truth? Clearly not, as Aristarchus and the atomists were correct. Later, Aristotle's philosophy of the soul was fortified by (Saint) Thomas Aquinas, according to William Anderson Gittens in his book *The Soul Of Culture Vol.1*, when talking about Aquinas' view as described in his work *Quaestiones Disputatae de Veritate*: "Concerning the human soul, his epistemological theory required that, since the knower becomes what he knows, the soul is definitely not corporeal — if it is corporeal when it knows what some corporeal thing is, that thing would come to be within it."[11]

This is wrong; my computer (which I believe to be "soulless" — or not... hmm) "knows" (it can list its contents) what my database is, but it only displays my database when I desire it. That is, it displays a temporary image of my database (which I might have entered as a material object, say a paper spread sheet that is scanned in). Information is non-corporeal, but it must have a corporeal substrate for storage and manipulation, whether on the original sheet of paper or its digital representation in semiconductor arrays. If the soul is the brain, it contains a representation of reality, not the reality itself — can we ever "know" a physical object? Only indirectly, with the limited senses and abilities we have. We do *know*, however, that our brain does all of what Aristotle attributed to the soul.

While it was humankind's first attempt to understand and control the mundane physical world, much of this early Greek "philosophy" was the application of fine reasoning to incorrect or inaccurate assumptions, axioms that simply have little or nothing to do with reality. The very idea of an immortal soul is an assumption without bases other than dreams, wishes and legends. People just like to hear that they'll live forever. For me, and I'm guessing most of my readers, my soul is what I mean when I say "I". Even if I lost memory of my prior life, having amnesia is not the same as dying. But when you read Plato and Aristotle on the soul, it is clear that Plato refers to an abstract heaven and an abstract soul, neither of which seem to have anything to do with "I" (in fact getting rid of that "I" is the Buddhist path to salvation; what happens after death was never discussed by the Buddha, Gautama Siddhartha).

The soul, as partly advocated by the Church and confirmed and strengthened by the Church fathers, has neither the authority of Jesus, in whose name the Church was created, nor the authority of knowledge and understanding of present-day science. The ancient Greeks knew very little, and much that they thought they knew was wrong. What Aristotle was really talking about are the properties of the brain, and there's no question that the brain is not immortal — even the cerebral cortex, the center of thought and reason, rots like all flesh.

All in all, the Greeks were the first to look at this world and try to understand it in terms of logic and laws, just as their geometry was based on axioms, connected by logic to explain the world of geometric shapes, a world Plato believed to exist in reality, a meta-world of ideal pure forms, of which our world is a shoddy approximation — where else could ideas come from? Answer: experiences, analogies and the ability to generalize (and often overgeneralize) them.

In sum, as the "classical" Greeks knew very little of the world, their fine reasoning was based on false axioms, and the results were never checked by a scientific method, as it had not been developed. When "Church fathers", like St. Aquinas, further demonstrated the non-corporeal nature of the soul (a modern equivalent would be the difference between "mind" and "brain"), the body became like a machine, animated by a ghost. How those immaterial fingers could control the switches of the body, and how that ghost could perceive light — though light would pass without disturbance through its invisible eyes — were questions that were never answered, but it was supposed the answers would come. However, we know the actions of the mind (though non-corporeal) reside in the actions of brain cells, which certainly die.

The Church took faulty human conclusions based on inaccurate and incorrect conceptions and made them dogma. A billion people today are still convinced by authorities of the Church, who feed them stories, and take their money to support a huge and wealthy bureaucracy. But we'll see before this book ends that though the Church's path to immortality through immortal souls is nothing but a fairy-tale based on misunderstanding, the scientific establishment followed the same path of incorporating untruths and fervently held wishes into an official "dogma" that has no more chance of giving us biological immortality (or rather, youth, for as long as we want) than the fictional immortal soul.

The ultimate answer to Descartes' mind-body dualism (his theory on the separation between the mind and the body) is that it is, as Gilbert Ryle (the British philosopher of the mind) explained, a fundamental mistake that cannot be corrected by minor alterations. Ryle refers to the mind-body duality as "the Ghost in the Machine". The idea that humans are composed of two substances (body and mind), as Ryle suggests in his essay *Descartes' Myth*,[12] is a categorical mistake, as body and mind are not "substances"; they are not the same kinds of thing, they belong to different categories — just as a "novel" and "writing" are two different categories, or "hand" and "washing", or "program" and "computer" (How much does that program weigh?), as I suggested earlier. "I have two hands and two washings" makes no sense, it's far worse than comparing apples to oranges. At least apples and oranges are in the category of fruits.

The medieval misunderstandings are still held by billions of people, who give their money to support a huge structure built to defend medieval thinking and medieval ignorance. However, human beings are (so far as we know) unique in our understanding that we will age and die, and so it is perhaps better to have hope that all the work and efforts put into obeying the generally moral religious laws will pay out as an afterlife, than accepting death as final and therefore knowing that any crime you can get away with will have no further consequences. Nevertheless, when I lived in Japan (for eight years), the average man in the street would tell you (and I am witness) that the Japanese don't believe in God, and yet crime is much lower, and civility much higher in Japan than it is in Christian nations, so I don't hold that fear of eternal punishment as a deterrent to crime; as I understand it, Mafia bosses go to Church every Sunday.

So, let's go back to a bit more biology, because it seems the appropriate place to look for a more scientific approach to extending life. We shall again have to call upon Feynman's principle — the principle by which all sciences move forward, which put into a reverse phraseology (because it's logically correct) is that you know the hypothesis you've chosen is wrong if it doesn't "work". This is called "rejecting the hypothesis"; in this infinite universe you can never know something is correct in every case, in every place, so you can never actually confirm a hypothesis by experiment, but any single instance where a prediction based on your hypothesis doesn't work is enough to "reject" that hypothesis and disqualify it forever without further testing.

Science is very conservative that way, yet there might be a lot of ways your principle (axiom, law, guess) is wrong and you still produce good results (as we saw in the case of Ptolemy's geocentric "universe" versus Copernicus' heliocentric universe). But the single instance of Galileo showing the phases of Venus (only made visible through his perfection of the telescope) was in itself sufficient to reject the hypothesis that Venus and the Sun circled the Earth.

Actually, Brahe came up with another model whereby the Sun still orbited the Earth, but the inner planets orbited the Sun. That theory could have also worked, but it would introduce further confusion. Assuming heliocentricity plus calculating the elliptical orbits of planets around the Sun at one focus of that orbit made everything clear and led to further progress, especially by Galileo and then Newton. There are two counterintuitive facts here; that better hypotheses don't necessarily give better results and that so-called Occam's Razor, the insistence that the simplest explanation is the best, isn't always the best way to proceed, as the Sun does not orbit the Earth every day.

4

Biology begins

Physical science, as it originated with Galileo and Newton, and even astronomy, appeared simple in principle; if a mere four laws can describe and predict the orbits of celestial bodies, then they are simple by definition. Of course, the physics of the time thought about mass points — a planet could simply be represented as a single point with a mass and velocity subjected to forces from the Sun and other planets, and using Newton's laws of motion and gravitation, their movements could be determined till the end of time (it was believed). However, extending physics to "extended bodies", like "real" planets, was for the future; the "mass point" and "forces" were all the physicist or astronomer needed. But biology was different — its complexity made it hard to even know where to begin; what were the equivalents of orbits and what were the "mass points" of biology? As a starting point, living things were known to be different, made alive by an *élan vital** peculiar to living things.

---

* *Élan vital*, "vital impetus" or "vital force", is a hypothetical explanation for evolution and development of organisms.

Though the discipline of *biology* only began in the 19th century, it had its roots, as much of science, in ancient Greek philosophy, whose enquiring spirit reached an apex (there were many) with Galen (Aelius Galenus) in the second century BCE. Galen was physician to the Roman Emperor Marcus Aurelius, and later to his son, Emperor Commodus (see the movie Gladiator to get an extremely biased view of Commodus, but historians mostly agree that he ended the Eastern Roman Empire as a force in the world).

Galen was an empiricist who wanted to study the human body to enhance his knowledge of medicine, and since human dissection and vivisection were made illegal in Rome in 150 BCE, he dissected and vivisected monkeys. Later I'll tell you about other experiments, so cruel it would be hard to imagine any "Ethics Committee" justifying them, yet some of them led to the discovery of a basis for mammalian biological immortality. Truthfully, Galen himself was discomforted by the human expressions on the faces of his monkeys and switched to other animals, mainly pigs. Galen wrote a treatise, apparently praising his own approach, titled appropriately *That the Best Physician Is Also a Philosopher*, as Galen was both physician and philosopher, and recognized as preeminent in both fields by none less than Marcus Aurelius himself.

The good part of Galen's investigation consisted of detailed dissections, with inferred functions from the varying part. Galen's anatomy survived up until the 13th century in Arab lands, when Ibn al-Nafis demonstrated the pulmonary circulation (the lung did not eat up the blood pumped to it, but it was recovered in the venous circulation),[13] and in Europe up until the 16th century, when Vesalius, a Flemish anatomist and physician, used the dissection of human cadavers (allowed in Christian Europe, but forbidden in Pagan Rome — not what I'd expect) to show that many of Galen's anatomical descriptions based on monkeys did not match those of humans.

Galen's philosophy was based on the Platonists and Aristotle's philosophy, including a three-part soul (an appetitive soul, a spirit soul ["spirit", as in the Hebrew "breath", meaning an animating force, not a "ghost"] and the rational part [that part which Aristotle believed immortal]). However, by experimentation, Galen believed he determined the location of these three "souls". The appetitive soul resided in the liver where it manufactured "dark" blood and sent it out to the organs where it was consumed. The "spiritual" soul resided in the heart, where it made "bright" blood that the heart pumped to the organs where it was consumed. And while Galen did distinguish between "bright" and "dark" blood, he never made the connection between the venous and arterial systems, so in both cases blood was produced under the direction of these two mortal souls (appetitive and spirit) in a one-way trip — the heart and liver made the blood, the other organs consumed it, then the heart and liver made more and the cycle continues. However, the rational soul was in the brain. The important part was not so much the knowledge gained, but the idea of "localization of function", the concept that different organs had different functions.

So here the Greek concept of *pneuma* was nearly identical to the Hebrew concept of *neshamah* or "breath" — that animating spirit which the stoics (the philosophical sect of which Marcus Aurelius was the most famous example) believed lay in the blood (remember, "the blood is the life of all living creatures"). However, Galen's medicine was based on the concepts of the Greek physician Hippocrates, who believed that health consisted in the balance of the four humors — blood, black bile, yellow bile and phlegm — so that bleeding to restore balance was part of the Hippocratic tradition (incidentally, donating blood is associated with increased longevity). The process of bleeding as a treatment for almost any condition continued well into the 20th century. I personally recall the set of cups my mother had for "cupping", a later variant of bloodletting.

Bloodletting would seem one way to clear the body of certain age-promoting molecules which appear to accumulate with age, like eotaxin (more later). By bleeding sufficient blood you could effectively dilute age-promoting (progeronic) factors present in the blood; since the body will readily make up the volume lost with water consumed, the progeronic factors should build up again, but regular bloodletting may lower apparent tissue age as it has been shown that dilution of such factors with partial replacement by a saline and albumin solution has such anti-aging effects,[14] except that young albumin was later shown to be anti-aging by itself, which puts dilution of aging factors as "cause" of rejuvenation in contention.[15]

Galen, however, was the end of the line for medicine until the Renaissance, when renewed interest in the real world, *empiricism*, and the appearance of ancient Latin and Greek texts (annotated by Islamic scholars) raised many new questions and the scientific method provided verified answers.

The Renaissance ("rebirth") was an age when a new philosophy came to supplant the ancient idea of the world as an organism and replaced it with the world as a mechanism. It is probably not coincidental that mechanisms too were being invented, as the scientific paradigm regarding biology is usually dependent on the technology of the age — such that in the 20th century, the computer became the model for the brain and vice versa (neural networks, AI, etc.).

It is reputed that the same Dutch spectacle-makers, Lippershey and Janssen, not only made the first telescopes, but the first microscopes as well. And, as with the telescope, Galileo improved on the microscope. After Galileo, it was Robert Hooke's microscope which most resembled the modern instrument (Newton was a great enemy of Hooke and it has been speculated that Newton's speech about seeing farther because he "stood on the shoulders of giants", was an indirect insult to Hooke, who was short). Hooke finally compiled all of his microscopic drawings of highly magnified common objects like fleas, lice, flowers and other plant parts into what we would now call a "coffee table" book, *Micrographia*, which provided a drawing of highly magnified cork showing the low-density bark to be composed of empty "little rooms" like the cells of a monastery, which is the origin of the word "cell", now used to identify what might be considered the "mass point" of life (and its progression through time as its "trajectory" or "orbit").

It wasn't professional scientists, however, who discovered the world of the cell, but an amateur, a hobbyist, Antonie van Leeuwenhoek. With the dedication of one who works for the love of it, Leeuwenhoek applied his passions to satisfying his own curiosity, but in the course of that, he discovered a previously invisible world of living things unknown and unsuspected that finally led to our current conception of the cell and much of our current biology.

Using a technique starting with the tiny thickenings at the bottom of blown glass, he was believed to have removed the rest of the glass leaving a tiny lens he sandwiched between brass plates (with holes for the lens). Though the microscopes developed by Hook and others were much more sophisticated, none had the magnifying power (more than 250 times) or resolution (0.001 mm) of Leeuwenhoek's marvelous lenses. This uneducated (with no formal education) dry-goods merchant made hundreds of microscopes (all had a system of screws such that the specimen is placed on one tip that can be raised or lowered relative to the lens).

Not only did Leeuwenhoek look at everything using his marvelous microscope; he also shared his findings with the British Royal society (even at a time the Netherlands and England were at war) and became a member thereof. He was the first person to describe spermatozoa, and when he looked at the water of his home city of Delft, he discovered the water he drank teemed with myriads of strange and graceful creatures we now call protozoa. When looking at the scrapings of his neighbor's very bad teeth and breathe, he was able to see the snake-like bacteria moving through the tartar and postulated that diseases might be caused by the invisible creatures he discovered.

I feel I have to also mention the great German researchers Matthias Schleiden and Theodor Schwann. Schleiden was a botanist, and Schwan a zoologist. Working in the same laboratory, they became friends and collaborators in the *cell theory*, that proposed, based on their evidence, that all living creatures are composed of cells. The original "cells" of cork were now filled with fluid; plant cells had thick cell walls of cellulose, as in the cork, but were filled with liquids or gels. Animals tissues too, when examined microscopically, were also composed of cells, but these had no cell walls — a thin membrane could be inferred to contain its liquid contents.

So, now we had our biological equivalent of a mass point, the cell. But where did cells come from? Originally, these "cells" were thought to be a liquid called *blastema*, as cells, like yeast, produced bubbles by the fermentative processing of foodstuff. But when the famous physician Rudolf Virchow announced his famous dictum, "Omnis cellula e cellula" (all cells come from pre-existing cells), it was accepted by biologists (although Robert Remak had made the same proposal before, an eminent authority such as Virchow stating it gave it credence). So the question of where our cells come from is that all we are is a clone of a single fertilized egg cell.

We're going to take a closer look at the cell, seeing that this "point mass" of biology has definite extent, then look at its life-trajectory, and take a close look at single cells that are at the same time independent living organisms — the "infusoria" that so fascinated Leeuwenhoek (protozoa) — and see why, how and when death comes to life.

The Eukaryotic Cell

No one quite knows what the Last Eukaryotic Common Ancestor (LECA) looked like or how it came into being (not to be confused with LUCA, the Last Universal Common Ancestor), but our recent ability to readily sequence the DNAs of any organism has allowed us to follow the ancestry of animals and to produce a new classification of living things based on their genomes. One example is shown in Figure 3.

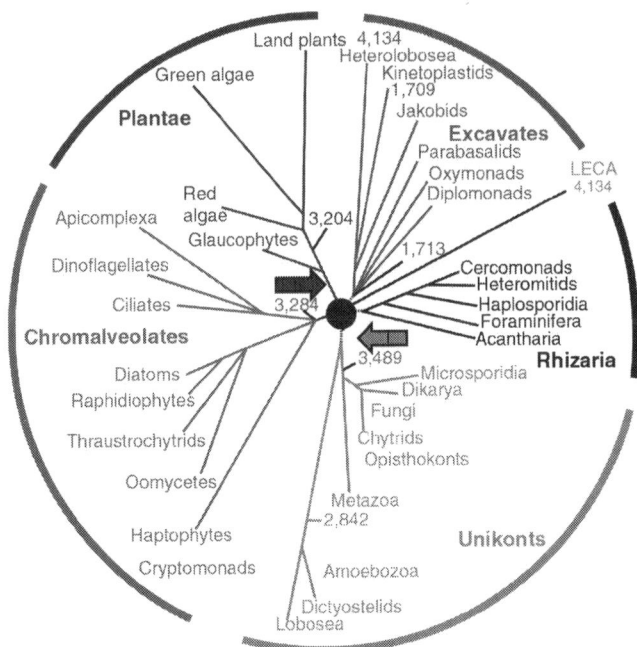

Figure 3: Evolution of the eukaryotes. The relationship between the five eukaryotic supergroups — Excavates, Rhizaria, Unikonts, Chromalveolates and Plantae — is shown as a star phylogeny with LECA placed in the center.[16]

Now, how do we know about our far-distant ancestors after so much time? Living things are conservative in the best way — they keep what's good, and (try to) get rid of what's bad. And the genes that they retain are retained often through evolutionary time. The genetic system that controls lifespan in the primitive roundworm *Caenorhabditis elegans* controls aspects of lifespan, particularly cellular responses to the environment (including signaling), in us. The insulin-mTORC1 pathway consists of homologous proteins and microRNA families in both roundworms and humans. When genes are different but similar enough to have had a common origin, we call them "homologues". When those genes work together to perform a function (such as the insulin-mTOR system in regulating growth vs. repair), they are called a Gene Regulatory Network (GRN). As we'll see, the GRNs that regulate the activities and lifespan of the roundworm *C. elegans* are homologues to our own.

When we look at the genes that members of the five eukaryotic supergroups share, we find that the last common ancestor to all eukaryotes probably had about 4,500 genes, pretty much the same as modern unicellular creatures, and when we look at the composition of those genes, many are of prokaryote origin; about three-quarters of a eukaryote's prokaryotic genes come from bacteria and about one quarter from archaeal DNA.

The DNA of archaea is like our DNA, with meaningful stretches coding for an amino acid sequence (a protein), and the complete sequence of the functional protein is interrupted by intervening and sometimes meaningless stretches of DNA (introns) — "meaningless" in the sense of not providing the correct sequence of the chain of amino acids proteins are made of. So, as those "meaningless" introns have to be removed, they are marked with their own "code" (other than the genetic code) for their removal by "splicing factors", and the meaningful parts, exons (for "expressed sequences") have to be stitched together. The entire complex operation is called *splicing* (in analogy with splicing a film, or audio tape, to remove unwanted parts), before the protein code carried by the RNA, transcribed from the archaeal or eukaryotic DNA, can be translated into protein.

It's not surprising that the control of information processing and transcription of RNA from DNA, the splicing of that RNA, and the translation of that RNA derived from archaeal genes, with the same intron/exon structure as eukaryotic genes, while the more plentiful bacterial genes were more concerned with metabolism. The major innovation of the eukaryotic cell was its complex composition, including membrane-bound organelles, and membranes forming tubes and bubbles within the cell — something unheard of in bacteria except for the few cases where one bacterium is an endoparasite of another. Three of these organelles are what makes the eukaryotic cell able to accomplish feats the prokaryotes could never achieve and all three are based on double layers of bilipid membranes; the nucleus, the mitochondrion and the chloroplast.

The nucleus contains the genetic codes, and epigenetic codes and mechanisms that will determine the characteristics of the organism. The mitochondrion gives the cell a vastly more efficient engine for the manufacture of the universal cellular fuel molecule, ATP, but its "exhaust" causes damage. The mitochondrion boosts the two molecules of ATP each molecule of glucose (grape sugar) can produce in a cell without mitochondria to an average of 32 molecules of ATP. The mitochondrion takes the high potential energy electrons donated by the molecules NADH and $FADH_2$ down its electron transport chain (ETC) to combine them with atmospheric oxygen molecules dissolved in the plasma.

By doing this, there is a loss of electron potential energy from its input as high potential energy NADH (hydride ions attached to NAD^+) to its output as low potential energy water and this difference of electric potential energy is used to generate ATP molecules from ADP and inorganic (energy poor) phosphate.

Every cell with mitochondria has a nucleus, and similarly, every cell with chloroplasts has mitochondria. The chloroplast, in combination with the mitochondrion, allows Plantae to make its own food from atmospheric carbon dioxide, water and sunlight (plus phosphates, nitrates, etc.). Solar energy was already a good idea more than three billion years ago. The chloroplast reduces (adds electrons, or in reality, hydrogen atoms to) carbon dioxide to make carbohydrate foods, and the mitochondria use this food (fuel) to produce ATP, which runs all the machinery of the cell. In particular, plants produce atmospheric oxygen, a very reactive molecule that must be constantly replaced by Plantae species as it has a short free half-life. It is this atmospheric oxygen that was believed to result in the first Great Extinction event, when formerly anaerobic organisms were exposed to it, but it also gave living things that could use this deadly chemical (O_2) the opportunity of using the excess energy to do more than just live and reproduce their kind. I will talk more about both of these organelles at length, but not about plants; though they too age and die, saving them from that fate is not our concern, nor is it theirs.

We know that the mitochondrion resulted from an endosymbiotic association of a large cell or colony of cells — with both bacterial and archaeal genes — with a bacterium of the gram-negative group of bacteria called alpha-proteobacteria that then became our mitochondrion. This is well established. The advantage provided by oxygen was enormous, and advanced eukaryotic life depends on the presence of atmospheric oxygen as a final low potential energy repository for the high energy electrons from organic compounds.

A molecule such as glucose (a simple sugar found in fruits) yields nearly twenty-fold the energy with oxidative phosphorylation in mitochondria (the process of making ATP inside the mitochondria) than if the mitochondria are removed or inactivated, thus giving cells more energy than they needed for survival and reproduction, but like electricity, it could be used for any number of purposes. And that glucose molecule was provided by plants who harnessed cyanobacteria to provide the solar energy the world of life depends on (with the small exceptions of organisms that live deep beneath the ocean, dependent on the high-energy compounds carried by thermal vents to energize them).

The double membrane surrounding the nucleus is indicative of a gram-negative bacterium. The nuclear pore complex that discriminates what may enter or leave the nucleus is also of bacterial origin (with some very randomly repeated peptide sequences that are not found in bacteria or archaea). It has been proposed that the incorporation of the alpha-proteobacteria as mitochondria may have resulted in the release of the self-splicing introns the alpha-proteobacterial DNA contains in abundance. These bits of DNA could interfere with transcription, necessitating a nuclear envelope to isolate transcription (but necessity doesn't necessarily make it happen).

Heterotrophic organisms (feeding on others) rather than the plant Kingdom is our present concern, but it often turns out that research in plants discovers mechanisms before they are discovered in animals, and while I will not devote our attention to plants, we should regard the definition of plant senescence as given by Hafsi Miloud and Guendouz Ali: "Senescence in plants is defined as the age-dependent programmed degradation and degeneration process of cells, organs or the entire organism, leading to death."[17]

We're all familiar with the colors of plant death, as leaves stop making chlorophyll to ready themselves for death as winter arrives in temperate climates. As the leaves die, an abscission layer forms between the leaf stem and its supporting vasculature, allowing the dead leaf to fall to the ground replenishing the soil. And most of our cereals — oats, wheat, rice, etc. — die before they're harvested. I remember watching the rice paddies near my home in Japan turning golden, not realizing that like fallen leaves, gold was the color of death.

The presence of annual, biennial and perennial plants (including trees that live from tens to thousands of years) shows that the same, age-dependent programed degeneration that we call "aging" in animals also occurs in plants as a normal part of their developmental program, except plants measure age by season, at least in temperate climes. More importantly, the very same, genetic, age-dependent programmed degeneration leading to death is responsible for our own aging and death — something that was hard to face. However, as it's often the case, squarely facing the problem provided the solution.

Protozoa

This is an unofficial term for what Leeuwenhoek called "animalcules" (little animals), and indeed many have animal characteristics, as they are mobile heterotrophic hunters, though some do have endosymbiotic cyanobacteria which provide them energy. Indeed, the animal called *Astasia* is identical to the plant *Euglena* except for the chloroplasts. "Cure" the *Euglena* of its chloroplasts (using a DNA intercalating agent), and it becomes the heterotroph *Astasia*.

Really, though, if you look at the five supergroups in Figure 3, you'll see that there are members of the protozoa (each group has a monophyletic origin, i.e., originated from a single ancestral species) belonging to each group. But for our purposes, we'll divide them up somewhat differently.

The immortal cells have no form of sexual reproduction, so their only means of reproduction is vegetative — they merely reproduce themselves with incredible accuracy such that these animals have been alive for billions of years! But in this case, it's the immortality of cell lines, not individual cells. Of course, in the case of these asexual protozoans, the members of a particular species are basically fungible, like other fungible items (like dollar bills, one can be exchanged for another). Mutations and DNA damage and misrepair do occur and changes accumulate, which adapt these organisms to their environment — so that evolution takes place, but slowly, as "Muller's ratchet" limits change.

Herman Muller's "ratchet" is the concept that asexually reproducing living things having a static arrangement of genes on their chromosomes (representing a single "linkage group") will be irreversibly damaged as their offspring will receive the same damages their parents had, and can only add to them. The sexual process of reproduction includes what is called genetic recombination, a process whereby defective genes may be removed from at least some of the progeny (this is often given as an advantage of sexual reproduction). Interestingly, in experiments with the bacterium *Escherichia coli* and the symmetrically dividing yeast *Schizosaccharomyces pombe*, the cells were immortal, and showed no signs of aging (measured by division rate), but did show aging when raised in a stressed environment. Might we suppose there's just "so much" these organisms can handle? A limit to what they can fight off?

Love and death among the protozoa

Let's be honest, we're not talking about love here; love seems to be restricted to mammals and perhaps birds — we're talking about sexual reproduction.

Sexual reproduction, in principle, started earlier than even the eukaryotes; it started as a bacterial disease. A plasmid is a mobile genetic vector, a small circle of DNA that contains several dozen genes, and so, it is much smaller than the bacterial DNA ("chromosome"), which is also a circle. When a sex-inducing plasmid enters a cell, it prevents further entry of similar plasmids, so every cell contains at most one of these plasmids. Some of these plasmids can become part of their host DNA, integrating seamlessly into the host's (bacterium's) DNA circle as a linear molecule, and the bacterium passes those "integrated" viruses on like its own genes — the enzyme that duplicates DNA, *DNA polymerase*, can't tell the difference, DNA is DNA.

In the well-studied bacterium *E. coli* (nothing to be scared of; about 1/6 of your poop is *E. coli*), that plasmid is called an F-plasmid. When it enters a cell of *E. coli*, it changes that bacterium, as it now has a "sex", so the bacterium is now called an F^+ bacterium. Some of the F-plasmid's genes are used to form a penis-like projection called a pilus (though DNA doesn't travel through it) that attaches it to any F^- bacterium (a bacterium that does not have an F-plasmid) it finds; it nicks itself, and then duplicates a single strand of its own DNA and sends that single-stranded copy of its own DNA into the F^- cell, which now becomes F^+ (you might consider F^+ male and F^- female, by analogy, but in this case males turn females into males — or non-carriers into carriers of F-plasmids). But if that F-plasmid is inserted (integrated) into the host's (F^+ *E. coli*) DNA, at the time this whole process occurs, when the pilus connects the F^+ with an F^- bacterium, instead of simply producing a single-stranded copy of its own F-plasmid DNA to transfer into the F^- cell, it transfers a single-stranded copy of both its F-plasmid DNA and the whole F^+ bacterial chromosome! If it has time, the total transfer takes hours.

Now, it's not really sex, as there are no gametes, and a totally new organism isn't formed, but genes get transferred from one bacterium (the F^+) to another (the F^-) of the same species and sometimes those foreign genes can be quite useful to the bacterium (there are several human diseases that are due to the genes carried by plasmids, so lethal bacteria would be harmless without them). Even *E. coli* can produce enterotoxins (poisons that damage the gut) when infected with the "right" plasmid, and that induces the horrified reaction to the name *E. coli.*

So, let's not call it sex; it's normally referred to as *horizontal gene transfer.* Bacteria can also pick DNA up out of their surroundings and add it to their own DNA, which is called *transformation,* or have small pieces of foreign DNA carried into them by bacterial viruses that can integrate into their DNA, which is called *transduction* — but as these modes don't involve the direct contact between individuals they don't resemble sex. Maybe it will resemble sex in the future, as in Isaac Asimov's *The Naked Sun* science fiction novel, in which people in far-off colony worlds are appalled to actually be with someone in the flesh, instead of the 25th century's equivalent of Facetime (will someday sexual reproduction be done by Space Mail sperm delivery?).[18]

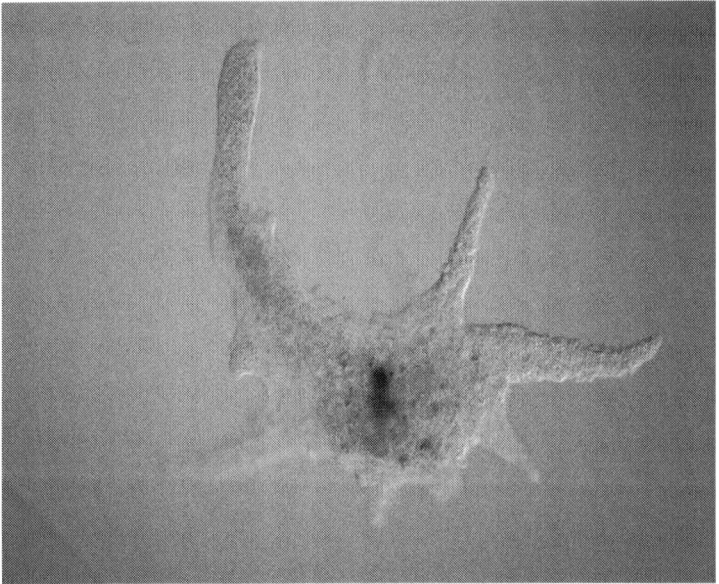

Figure 4: Amoeba protozoan *Chaos carolinense*. Image by dr. Tsukii Yuuji, CC BY-SA 2.5 https://creativecommons.org/licenses/by-sa/2.5, via Wikimedia Commons.

Among sexually reproducing creatures, every chromosome contains genes with different functions, for example, genes for eye color, a gene for making a hemoglobin molecule, or genes determining cell defenses. All sexually reproducing organisms have two of each kind of chromosome; they carry the same genes — for eye color, for example — but as there are two of them, though both carry eye-color genes, they may carry genes for different eye colors (alleles), such that if a person inherits one gene for blue eyes and one for brown, hazel or gray eyes would be possible phenotypes. Organisms that contain two of each variety of chromosomes are called *diploids*, and only these can have sexual reproduction. Animals that have only one of each type are called *haploids*.

So, among the simpler unicellular animals, the amoeba pictured in Figure 4 is a good example. These unicellular animals are immortal, as are many other unicellular protozoans. Figure 5 shows a tetrahymena, still a unicellular protozoan, but there's a difference: it is mortal.

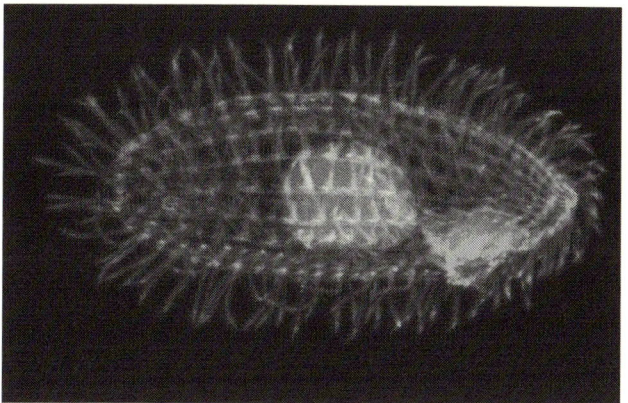

Figure 5: Ciliate protozoan *Tetrahymena thermophila*. Image by Richard Robinson.[19]

And that's the distinction I want to make between protozoans — not which supergroup they belong to, but whether they are mortal or immortal, to us the all-important distinction. In the illustration in Figure 5, the large light gray central area inside de cell is the *macronucleus*, but if this is a mortal strain (about 25% of tetrahymena caught in the wild are immortal), there is another much smaller nucleus called *micronucleus* which is transcriptionally inactive except during sexual union (called *conjugation* in protozoans and bacteria). The micronucleus is a germ nucleus, the equivalent of the germ tissues that form gametes, sperm and ova, in higher animals. It contains the full diploid genome of the ciliate — in the case of *T. thermophila*, depicted in Figure 5, five chromosomes.

During normal vegetative cell divisions, the tetrahymena undergoes mitosis, when exact copies are made and distributed to the two conjugants (the process in the macronucleus is much more haphazard, the divisions are non-mitotic and inexact). During conjugation, however, the micronucleus undergoes meiosis, forming four haploid gamete nuclei, much like the cells that produce our sperm or egg cells. Three of these gamete nuclei will be eliminated and the chosen nucleus will undergo mitosis, and the conjugants will reciprocally pass one of each pair of gametic haploid nuclei to the other, which will then join. So now, both cells are different from either "parent", in terms of their micronuclei; they are entirely new creatures but identical to each other (see Figure 6).

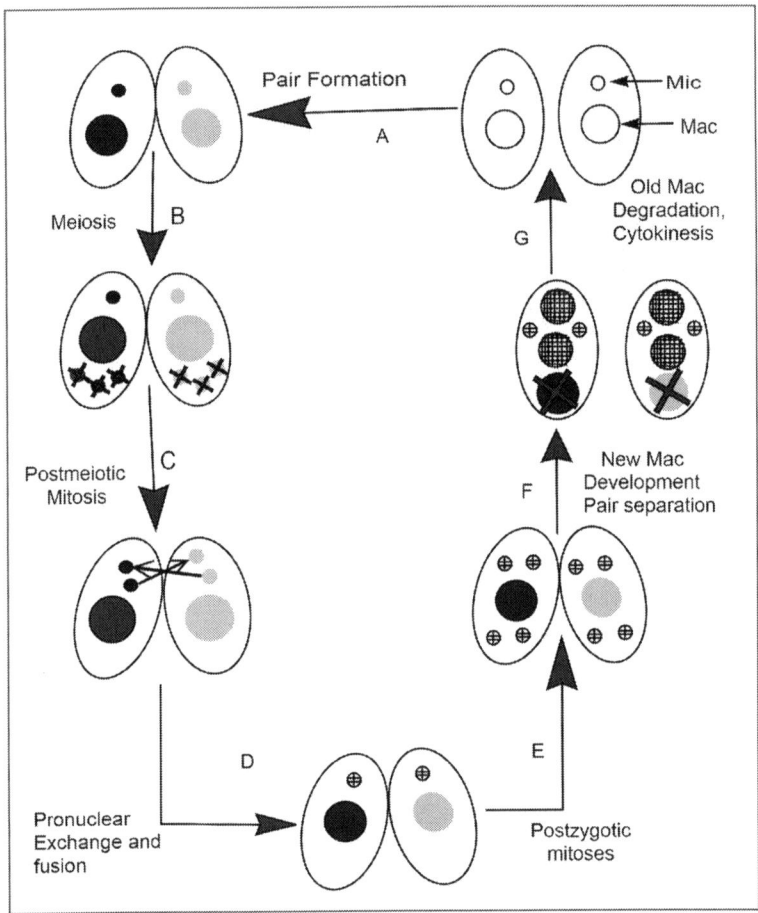

Figure 6: Tetrahymena conjugation. Image by Chaya5260, CC BY-SA 3.0 https://creativecommons.org/licenses/by-sa/3.0, via Wikimedia Commons.

In the process, the macronucleus that carries out the normal functioning of this organism is destroyed and a new one is created from the genes of one of the new hybrid micronuclei. One of the micronuclei differentiates into a macronucleus, with extensive rearrangement of genes and elimination of thousands of micronucleus-specific DNA sequences, finally breaking the five tetrahymenal chromosomes into about 200 pieces, each "piece" duplicated about 45 times and all with telomeres attached (a process that involves the DNA repair enzyme Ku80), after the removal of micronucleus-specific DNA sequences in the newly formed macronucleus.

The act of sexual reproduction (or autogamy — "self-sex") resets the "age-clock" of the organisms to zero; they now have the potential to live for whatever the maximum number of cell divisions are for that species. It is the macronucleus (Mac) that is transcriptionally active and determines the cell's phenotype (the way it looks, acts, behaves). The micronucleus (Mic) is transcriptionally active only during conjugation. The macronucleus is the actual somatic nucleus of the ciliate as it contains all of the genes needed for the animal to function. This majority population of sexually reproducing tetrahymena forms the mortal tetrahymena clones. This sexual process, conjugation, is a bit more complex than I led on, but reaches the conclusion shown in Figure 6.

The lifecycle of a ciliate protozoan

However, when I picked tetrahymena to discuss, I did not choose it randomly. Most ciliates are mortal, sexually reproducing unicells. But unlike other ciliates, about 50% of tetrahymena species lack a micronucleus (they are *amicronucleate*) and are therefore asexually reproducing — and immortal! In the laboratory, however, removal of the micronucleus does not result in viable amicronucleates, except in one known case in which micronuclear sequences, normally removed when forming the macronucleus, are retained in this viable, laboratory-produced amicronucleate *Tetrahymena thermophila* clone. One interesting thing about this amicronucleate is that it does not attempt to mate, though it appears to be genetically capable of mating as it has the mating type (MAT) genes needed.[20] This could be done by a process called nuclear exclusion — if an amicronucleate mates with a micronucleus-containing animal, the micronucleus will serve both animals.

One hypothesis for why these mutants do not mate is that they are in a permanent state of immaturity. Normally, laboratory strains of *T. thermophila* require from 40 to 60 fissions following conjugation before they can mate, and wild *T. thermophila* may be immature for up to 120 fissions following conjugation. It is perhaps surprising to see that this tiny simple creature has life stages, including an immature, sexless phase, a reproductive phase and a senescent phase, much as we do.

During the course of its life as a "normal" micronucleated ciliate, tetrahymena's macronucleus decays in functionality as the organism ages (where "age" is calculated as the numbers of cell divisions since fertilization), eventually needing replacement by mating with another tetrahymena in order to rebuild a functional macronucleus. After mating, neither of the original organisms' genomes survive intact, only the hybrid genome formed by sexual union, and hence a hybrid macronucleus derived from one of the micronuclei differentiates from these mixed genes.

Arguably, as the macronucleus determines the phenotype, the original organisms (which unlike the immortal protozoans are not fungible) are gone; two new (though identical) animals are formed following conjugation, so it could be said that both of the original ciliates died in the process of sexual reproduction, which is equivalent to the semelparous mode of reproduction, where annual plants, octopuses and most famously salmon, 17-year cicadas and mayflies, live long lives in immature stages only to reproduce and then almost immediately die. But the tetrahymena clones survive — the strain survives at the cost of individuals.

At conjugation, the old macronucleus degenerates in the micronucleus-containing tetrahymena unless conjugation occurs to reform a "fresh" macronucleus (and a new animal). Yet, amicronuclear species are immortal, with some clades estimated to be tens of millions of years old, so clearly the same processes that cause the ultimate wearing-out of the macronucleus in tetrahymena with micronuclei don't occur in amicronucleate tetrahymena!

So, can it be that preventing the progression to sexual maturity, which might require micronuclear involvement, halts the accumulation of genomic damages that mark the normally-aging macronuclei of micronucleus-containing tetrahymena? Might the micronucleus control development in tetrahymena, or might the micronucleus suppress the repair of macronuclear DNA damage during aging? It is known, for example, that secretions of the egg cells of the roundworm *Caenorhabditis elegans*[*] de-

---

[*] *C. elegans*, a very elegant and transparent nematode (round worm) used in many aging studies as it only lives a couple of weeks.

crease the lifespan of the worm, such that if those ova are removed, the worm lives significantly longer.

A partial answer to these questions was given by the work of Joan Smith-Sonneborn back in the 1970s with her study of another ciliate, the well-known genus *Paramecium*, species *tetraurelia*, pictured in Figure 7. This species has an immature period, a sexual maturity period, a senescence period, and finally it dies, unless the paramecium mates with another of the appropriate mating type (there are nine "sexes" in *P. tetraurelia*, that determine who can mate with whom), or with itself — a process called autogamy, with the paramecium's micronucleus undergoing meiosis and self-fertilization that results in a newly formed macronucleus. Autogamy is like conjugation in that it creates a new, fresh macronucleus, but it's unlike conjugation in that a missing or mutated gene has no chance of being complemented by a good working copy of that gene from its mate, so fatal recessive mutations, losses or lesions in micronuclear chromosomal DNA remain fatal and result in deaths following autogamy.

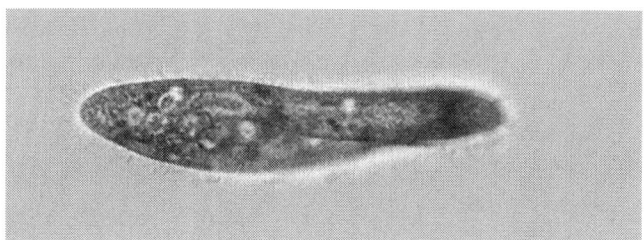

Figure 7: *Paramecium tetraurelia* (approximately 100 μm long, barely visible to the naked eye). Picture by DavidpBowman, CC BY-SA 4.0 https://creativecommons.org/licenses/by-sa/4.0, via Wikimedia Commons.

Let us not miss what we see here (and in the case of the ciliates in general): these organisms age, not in terms of years but in terms of the number of cell divisions, and furthermore have a life divided into life-stages according to age (in cell divisions). The life of paramecia in culture is, according to Leonard Hayflick (the discoverer of the sad fact that cultured human fibroblast cells have a finite lifetime in terms of their permitted number of cell divisions — a highly variable lifetime, but with an upper limit), very similar to human cells in culture. But there's a difference, isn't there? Human fibroblasts in culture are fungible, replaceable parts; tetrahymena in culture, at least those with micronuclei, are organisms, independent and not fungible. They are given no choice, they die (lose their identities) by sexual reproduction, or die by lack of sexual reproduction through loss of macronuclear functionality — though they might have lived longer if they allowed senescence to kill them, as at least half of all tetrahymena species have micronuclei and all paramecia do.

The loss of immortality to sexual reproduction was a winning trend. The overwhelming majority of multicellular animals and plants adopted it. However, there are still animals like the common cnidarian, such as jellyfish, sea anemones, corals, the *Hydra vulgaris* (*vulgaris* meaning "common"), flatworms and sponges that are immortal or nearly so — a common sponge has been estimated to be 11,000 years old.[21] The voluntary loss of identity through sexual reproduction is weighed against the betterment of the species, in the case of tetrahymena species, and both ways of reproduction seem to work well for this organism as half of all species are amicronucleate without a clear micronucleus-possessing ancestor.

However, tetrahymena are the exceptions among ciliates; asexually reproducing ciliates are rare, and in the case of paramecia, which I want to discuss next, there are no amicronuclear immortals, and lifespan is given as a range in the number of cell divisions following fertilization (but with a species maximum as we find in higher organisms). The lives of the sexually reproducing, mortal individuals are divided similarly into a sexually immature stage, a sexually mature stage and a senescent stage if sexual union doesn't take place — very much like us, though typically sex doesn't end our lives.

Now, getting back to Joan Smith-Sonneborn's experiments on aging in *Paramecium tetraurelia*, the first thing she did was to notice that if damage occurs to micronuclear DNA, autogamy will result in the deaths of both conjugants. One final result of the study is shown in Figure 8.

In Figure 8, it is clear that from fertilization until about the 60th generation following fertilization, the viability of the cells following autogamy remains close to 100%. And from then onwards, a linear decrease in viability after autogamy occurs such that by the 220th cell division, according to work by Sonneborn and Schneller,[22] the survival rate is zero — as this number of cell divisions is the species maximum age (in terms of cell divisions) between sexual reproduction or autogamy. Furthermore, an increased death at autogamy after UV irradiation above that of age-matched controls showed that "dark repair" declined with age.[21] "Dark repair" was a term (used in that age of ignorance of DNA lesions and repair processes) that included all DNA repair processes that did not involve light. Huh? DNA repair processes that used light?

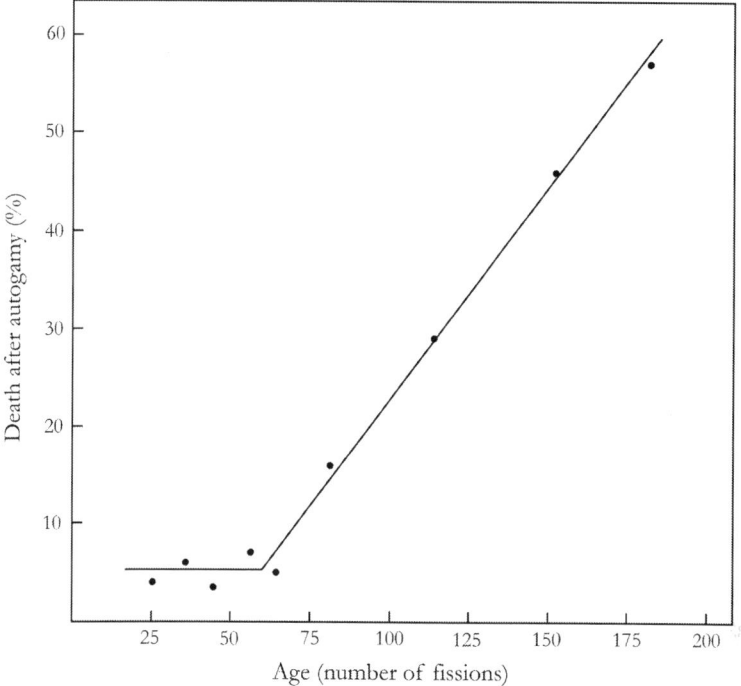

Figure 8: Death after autogamy as a function of age in *Paramecium tetraurelia*. Redrawn.[23] Autogamy means "self-mating", but like mating, it results in resetting the clock of the four emerging individuals. The death rate rises linearly with the number of cell divisions of *Paramecium tetraurelia*, past about 60 cell divisions. Does this correspond to an immature period?

It's simple, really. When you irradiate DNA with ultraviolet light, you cause adjacent bases of a certain kind (*pyrimidines*, cytosine and/or thymine — the other kinds are *purines*, adenine and guanine) to join together to form what is called a dimer (two molecules chemically fastened together), as it is shown in Figure 9.

Those "pyrimidine dimers" are the main sort of damages produced by irradiating DNA (in an organism) with UV radiation, but there are many other kinds of damages (I actually characterized some of them early in my career). But here's the thing — simple blue light can "undimerize" these pyrimidine dimers by itself. It is usually helped along by an enzyme called *photoreactivating enzyme* that speeds up the reaction (but still requires blue light).

According to Rodermeli and Smith-Sonneborn, UV irradiation increased the death rate of *Paramecium tetraurelia* after autogamy, but was it really DNA damage that caused the increase? Yes, because when the paramecia were exposed to photoreactivating blue light after UV irradiation — a process shown to remove pyrimidine dimers, and which should have removed most DNA lesions — the killing effect of UV irradiation prior to autogamy disappeared.[21]

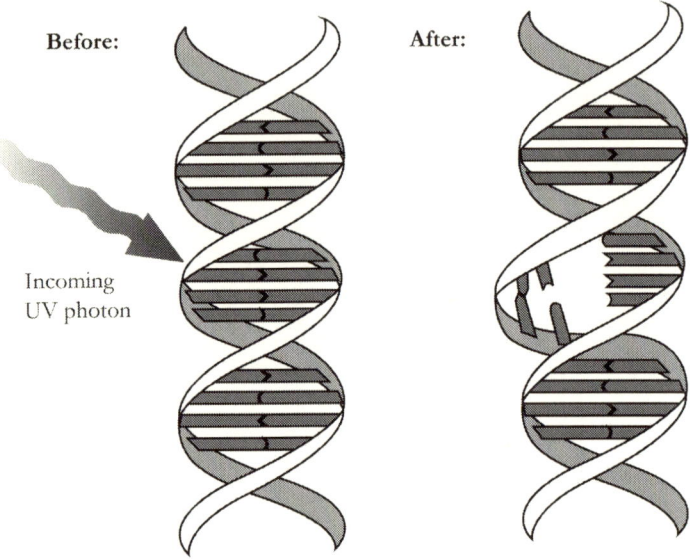

Figure 9: Effect of UV light in the formation of the thymine dimer.

Several conclusions and hypotheses were made by the authors,[21] including:
1. Aging (in terms of the generation of lethal micronuclear DNA damage by UV light) began at 50 to 80 cell divisions after fertilization (does this mean perfect repair in the immature period?)
2. A. Aging in the micronucleus resulted from an old cytoplasm, as young nuclei placed in the same old cytoplasm did not survive.

   OR

   B. Since lethal DNA damage or mutations result from both the generation of damage and its repair, "loss of repair in aged cells could account for an abrupt increase in mutations at a certain age, i.e., when the mutation rate exceeded repair capacity".

The 2A hypothesis was the result of the work of Joan's father-in-law, the famous founder of experimental protozoology, Tracy Sonneborn, but the second supposition, 2B, which resulted from the experiments Joan

Smith-Sonneborn performed herself, matches similar results in the cells of "higher" animals. Furthermore, that aging occurs when the DNA damage rate exceeds the innate capacity to repair that damage would explain the results obtained in the "immortal", symmetrically dividing cells of *E. coli* and *S. pombe* mentioned earlier. These organisms don't age in a friendly environment but only in a stressful one. In this case, the "environment" raises the stress levels, and damage-repair induction in the immortal cells is limited, and so eventually exceeds their capacity to repair that damage.

On the other hand, this abrupt increase in the ability of UV radiation to cause lethal micronuclear changes after about 70 to 80 cell divisions continues with age, and Smith-Sonneborn attributes that to the age-related decrease in error-free repair, rather than to a cytoplasmic factor — and yet these two explanations are not mutually exclusive; there could very well be cytoplasmic factors, possibly produced by the macronucleus or by the purportedly inactive micronucleus itself, that might turn down or turn off the production of repair enzymes. Smith-Sonneborn's own explanation for the experiment was "The simplest explanation of the above study would be that stochastic 'hits' mutate the micronucleus and the organism is programmed to lose error-free repair."[21] There's no shying away from "programmed" or "developmental" aging with this organism.

These are suppositions that make sense if it's really the case that aging is, in part, the result of UV light-induced DNA damage. But how can we really know that? We might return to Feynman's admonition that if you really understand a phenomenon, you can "build" it. So, if it is the case the UV-induced lesions are a cause of aging, then if you were to remove those lesions by some means, you should at least slow or possibly reverse the aging process.

The means of removing these lesions was already apparent from Smith-Sonneborn's first set of experiments; photo-reactivating, blue light. It was found that when paramecia were irradiated with UV, they had a shorter clonal lifespan (the number of divisions until the likelihood of successful vegetative [asexual] reproduction was zero). However, when UV irradiation was followed by a photo-reactivating light treatment (and not the other way around), there was an observable increase in their lifespans in terms of cell divisions and "age-dependent" survival rates, and if the process was repeated, an even more significant rejuvenation of the animal, or extension of its lifespan (they are different), occurred (see Figure 10).[24]

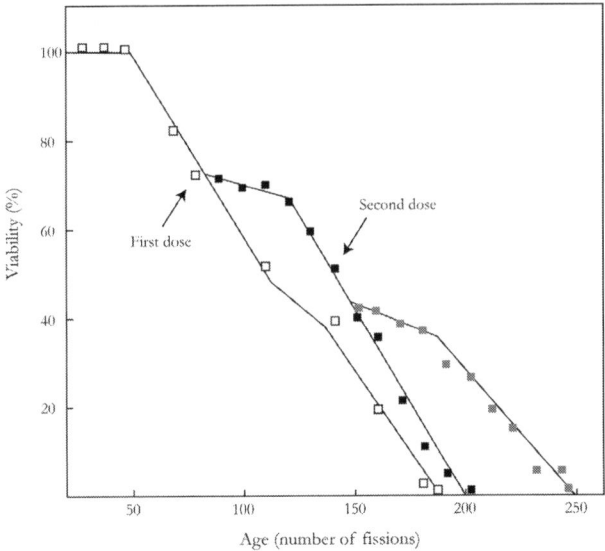

Figure 10: Lifespan in number of fissions of *Paramecium tetraurelia* after two doses of UV irradiation plus reactivating light treatment. Redrawn.[23]

In Figure 10, we can see that the untreated control clone (white squares) has a maximum "lifespan" of around 180 cell divisions, while a single UV/photoactivation treatment given at 80 fissions increased that by about 10%, and a second treatment, given to a fraction of the treated group, extended lifespan to more than 240 fissions, an increase of 33%. For humans, who live an average of 80 years, the average lifespan would reach 107 years if the same were to work on humans.

Further lifespan extension (or rejuvenation) by additional treatments was not discussed. What was discussed, however, was that using the UV followed by photoactivation procedure on younger cells (less than 80 "divisions old") produced no effect on clonal lifespan, and more importantly, that when the treatment was reversed, such that the UV treatment followed photoactivation, it resulted in shorter clonal lifespans as expected (photoactivation alone did not produce a marked change in clonal longevity).

Of course, Smith-Sonneborn followed the reasoning of the times, and yet, it was made clear by her work, and the related work we discussed, that:

1) Mortal ciliates have a lifespan limited by maximum lifespan past when none live.

2) Lifespan is divided into life-stages marked, among other traits, by age-specific (cell-division-specific mortality rates) such that during the first 80 or so divisions (the number of divisions for *P. tetraurelia* to sexually mature), there is no aging and no need to induce repair enzymes, so UV/photoreactivation does not extend lifespan (we might stipulate perfect repair, since, as we saw in the first set of experiment, there is no increase in mortality following autogamy at those ages either). After that "immature" stage, which appears to be a period before aging processes begin (at least in terms of the reduction of "dark repair" capacity), there is a steady increase in the accumulation of damage which Smith-Sonneborn attributes to a decrease in error-free repair rather than an increase in damage.

Smith-Sonneborn's and my own interpretation is that the extensive and potentially lethal damages produced by UV-radiation induced the production of DNA repair enzymes, which once freed of their responsibility of removing the pyrimidine dimers (which were removed by photoreactivation), were then free to repair accumulated damage, and as Figure 10 shows, repair newly occurring damage (as the downward curve of survival is shallower following two UV/photoreactivation treatments because there is a slower growth of damage, much as in the immature, pre-sexual stages).

Note that the slope of the curve following the first UV/photoreactivation remains nearly level for around 50 fissions, indicating extensive repair of damages, such that if accumulated DNA damage is a major timer (and cause) of aging, the age of the paramecium was almost halved as a result of treatment (but there is a wide span of fission numbers to reach senescence as time progresses).

So, it would seem that ciliates, such as micronuclear tetrahymena and paramecia, have predetermined lifespans, and two modes of reproduction: the first one is a vegetative mode of reproduction and requires meiosis in the micronuclei, and a rude form of nuclear division in the macronucleus together with destructive changes in both nuclei, with each cell division past "immaturity" reducing the cell's ability to survive autogamy; the other one is sexual reproduction by conjugation in which the two ex-conjugants are now both genetically identical but each different from either parent — a new organism with its life starting anew, now allowed approximately 180 cell divisions each individual (in the case of *P. tetraurelia*).

And yet the amicronucleate form of tetrahymena species (which we know in one case to have micronuclear DNA sequences not normally present in macronuclei) are immortal and have only a primitive form of nuclear division but are apparently able to perform error-free repair of their DNA. All the genes necessary for conjugation occur in these amicronuclear tetrahymena species and the reason they don't conjugate must be that these genes are never activated — which is the observed effect of immaturity, a lack of lethal mutations or lesions that result in death after autogamy. We're talking about 50 to 80 fissions (cell divisions) with relatively error-free repair during this life stage in mortal, sexual tetrahymena (i.e., following fertilization and preceding sexual maturity), and in paramecia as well.

We see in this case, as we shall see in the lives of all land vertebrates (our selfish interest) and perhaps all bilaterians (and all ciliates?) that there is a constant ratio between age at sexual maturity and length of life (in number of fissions, day or years) post-maturity — and hence a constant ratio between age at sexual maturity and total lifespan. As the ratio is about $65/200 = 37.5\%$, the immature period is about $1/3$ of the total lifespan (in non-flying terrestrial mammals). After the immature period, the accumulation of potentially fatal mutations occurs at a seemingly steady rate, and this has been shown to be a product of the loss of repair in aging micronucleate containing, sexually reproducing organisms and not fungible cells. Nature's command is clear here: reproduce sexually and die (as the original organism will no longer exist, only its "offspring"), or asexually and simply die (as, at this point, the asexual reproduction will fail and the organism will die without producing an "offspring").

As I said, this is logically equivalent to a semelparous reproductive strategy, as in the salmon and the octopus, where reproduction equals death — a common theme in nature, as mayflies, that spend productive lives as larvae underwater, transform to their sexual form (without even the mouthparts required to take in nutrition), live for a day, mate, lay eggs and die. The ultimate "purpose" of their lives is reproduction.

Let's however turn from biology to history to explain how the combination of the Darwinian principle of natural selection and Mendelian genetics came to form the so-called "Modern Synthesis" that dominated classical evolutionary theory and was responsible for the modern and incorrect understanding of aging. It's interesting from a philosophical, psychological and political standpoint, as this theory represents a major statement in the battle of science versus religion, and I believe it to be wrong.

5

The origin of species

What Darwin (as well as Alfred North Wallace) wrote in 1859, now titled *On The Origin of Species by Means of Natural Selection*, was originally proposed to be titled *On The Origin of Species by Survival of The Fittest*, but Darwin thought that too radical. Nonetheless, it represents the simple syllogistic logic that those who contribute the most viable, reproductively capable offspring will have a greater share of the genetic composition of future generations. This assumes that populations are limited, an idea Darwin received from Malthus' famous work, which I principally know as *population increases geometrically (i.e., exponentially) and food supply increases arithmetically* — which would have been disastrous were it not for human science and ingenuity. There is also the assumption that variation will always exist in natural populations (much information was derived from correspondence with animal breeders) and that there was a "principle of divergence", whereby the most divergent members of a species will be less likely to share a common niche (habitat, nesting sites, food preferences), and so the less intra-specific competition there will be. Then, natural selection tends to favor divergence.

So, natural selection, or survival of the fittest, proposed a mechanism wherein natural variation caused the adaptation of a species to its environment — that would hone and refine a species, accentuating extremes, which is a way of selecting superior-to-normal individuals and eliminating inferior-to-normal individuals. But are we really talking about the creation of new species? Darwin himself stated in his preface to the last edition of *On The Origin of The Species* that he thought "natural selection" to only be one of the ways evolution worked, and was disappointed that this caution was not noted in the general enthusiasm for this idea of "survival of the fittest". Indeed, that idea was readily adopted by an enthusiastic public shaping the Eugenics Movement, "scientific" racism and Nazism. It gives authority to racist claims of superiority and inferiority to people ignorant of history or the workings of science.

Survival of the fittest is not a test of "superiority"; it is simply a criterion based on which individuals' progeny will have a larger share in the species' future. While the movie *South Pacific* told us that racism "has to be carefully taught", I believe evidence shows it is a basic part of human nature to be suspicious of those that don't look like you and the people around you. It is common at the tribal level for members to call themselves "the people" (in their language) and others as non-people, either inferior or evil. At this time in the USA and across Europe we can see it as White Supremacy — in the next century it may be Han Supremacy, especially if the USA turns towards superstition, corruption and rule by ignorant mobs led by preachers, while China continues to pursue science.

As it happens, science sometimes becomes sidetracked by common-sense assumptions and false assumptions by authoritative figures and even by politics. I believe that it is this later fantastic component that created the opposition between Darwinists and religious authorities that countered evolutionary theory, mainly Protestant, especially "Evangelical" churches. The Catholic and Episcopal Churches eventually came to accept the mountains of evidence of evolution with the caveat that it was God who controlled evolution — a so-called "teleological argument" used by the priest Teilhard de Chardin to ultimately convince the Church of it (at the peril of excommunication, which to believing Catholics is, literally, "a fate worse than death").

The argument is that God used evolution to create humanity — evolution was God's tool. Smart priests are more the rule than the exception; it is my belief that enforcing celibacy on Catholic clergy was a way of eliminating the brightest commoners from the breeding pool — whether deliberate or not, it should have had that effect. Of course, in the higher echelons of the medieval Church, having illegitimate children (the Medici family being an example) was not the exception.

However, there was still a battle between religion and science: according to the Bible, the entirety of creation was achieved in six days. All animals, birds and crawling things were made in those six days. Thus, academic arguments occurred on whether the "days" mentioned in the Bible are the 24-hour days of today, or millennia-long days, before the creation of the sun (which occurred on the third day of creation). It is the sun's rising in the east and setting in the west that is the measure of our days. What could constitute a day beyond the duration of the sun in the sky? Even just a few decades after Darwin's death, it was widely suspected that the Earth was millions of years old due to radioactive dating of strata.

So, apparently, was agreement or disagreement with the Bible what decided the correctness of a scientific theory? The real question is: why an early iron-age text should dominate our understanding of the universe as we know so much more than they did two thousand years ago? It was Galileo's contention that "God wrote the handbook of the Heavens and it is written in the language of mathematics." It was also his contention that "The Bible shows the way to go to Heaven, not the way the Heavens go." If we have to deny reality in order to accept a religion, then our very first act in that religion is a lie. Darwin and Darwinism themselves became a sort of religion, with strict rules based on authority. We will now leave Darwin (though he's worth spending time with) and see how his theory, restricted by devout adherents of individual selection, affected most modern theories of aging, as that is our concern. What does Darwinism say about the evolution of lifespan?

Factors that influence evolution

First, before we accept "natural selection" as *the* mechanism of evolution, let's consider additional factors that are more likely to create new species than to perfect existing ones. As Dr. David Neill (to be discussed) has noted, natural selection explains microevolution, but not the longer-term, species-level changes and changes in higher taxa (like genus, family, class and order). One part of Darwin's thinking that made him decide that evolution occurred by small steps over generations was the analogy between the varieties of exotic animals produced by animal breeders — like "fancy" pigeons diverging very far from the norm, but we should consider they were still recognizably pigeons.

Factors other than "natural selection" that influence evolution:

1. **Mass extinctions:** during the evolution of life on Earth there have been several mass extinctions that have determined the trajectory of life on Earth.
2. **Hopeful monsters:** single gene mutations with pleiotropic effects can have significant effects on populations.
3. **Genetic drift:** it's important in small populations, where random events can eliminate rare alleles, which might be common in larger populations. This is particularly evident when islands are populated by a paucity of species (such as the famous Galapagos Islands, where Darwin saw clear evidence of evolution's efforts to adapt a population to its environment). Founder populations are an extreme example, in which the founders of a geographically new population are so few and hence so limited in diversity, that their offspring will be genetically limited to the alleles (gene variants — for example, black eyes rather than blue) that the founders carried, which may not be representative of the larger populations from which they derived.
4. **Gene flow:** where genes are exchanged by populations that have different alleles from each other.
5. **Epigenetics (and the evo/devo approach):** when the "modern synthesis" of evolutionary theory and Mendelian genetics first originated (a reference to the title of a book, *Evolution: The Modern Synthesis* by Julian Huxley in 1942 — hence more "modern" than "art nouveau"), it nearly completely overlooked the relationship between embryology and development. The "embryological view point" that "ontogeny recapitulates phylogeny" — that is, that the development of the organism from zygote (fertilized ovum) to birth is a recap (as we say) of the development of the species from simpler ancestors — was shelved in favor of the mathematical studies of overly simplified systems, but was nonetheless true, so far as they went, like the studies of Dobzhansky in population genetics.

The principles of evo-devo will be discussed later, but for now let us just note that Nature isn't a wastrel, but keeps what works over the eons, keeping functional collections of genes and their products, such as multicomponent gene regulatory networks (GRNs), together over long stretches of evolutionary distance (in some cases — such as the control of cellular energetics by a system starting with the insulin receptor and ending with the mTOR or FOXO proteins dominating — from roundworms to humans). We also see that the same GRNs control cellular age progression from roundworms to humans.

6. **Group selection:** the first example of group selection was given by Darwin himself in his book *The Descent of Man*, where he writes: "If one man in a tribe (...) invented a new snare or weapon, the tribe would increase in number, spread, and supplant other tribes. In a tribe thus rendered more numerous there would always be a rather better chance of the birth of other superior and inventive members."[25] However, long before the 1960s, evolutionary theory no longer accepted group selection as an important mechanism of evolution, except in the case of social insects, in which "kin selection" was the acceptable mechanism to the author of *The Selfish Gene*, Richard Dawkins, whose thesis was that selection occurs not only at the level of the individual, but at the level of the gene. In this conception, humans are mere mules for carrying their genes (Dawkins was also the originator of the word and concept of a *meme* — a self-propagating mental entity). In fact, what seems to be representative of thinking at this time is his statement that "Group selection on the level of the species is flawed because it is difficult to see how selective pressures would be applied to competing/non-cooperating individuals."[26]

To me this is another example of "argument by lack of imagination". I believe that this fundamental restriction of "natural selection" to individual selection transformed aging science into an intellectual parlor game rather than a search for truth. It is clear that the goal of biology is to produce immortality, and so far, only religion can make that claim, though without evidence. When we throw out our preconceptions, the limits of our own minds, we can see that biology can achieve those goals religions claim, but with proof, abundant proof.

Evolutionary theory and the meaning of aging

In a very interesting proposal, David Neill wrote an essay on the *Evolution of lifespan*.[27] The very name of his essay should be nonsensical, as, according to the authorities who developed the modern aging theory — the "modern synthesis" — aging was not a trait available for natural selection to operate on. Now, why was this? Because aging, as it occurs during an animal's post-reproductive period, cannot be selected against — as it is post-reproductive, it should not influence an individual's contribution to the future. Dr. Neill's paper clearly presents the problem when quoting John Maynard-Smith regarding August Weismann's theory — a foremost biologist of his time, who, in his 1882's book *Über die Dauer des Lebens*, said that an aging process would be helpful in removing a parental generation to free up resources for their fitter offspring. "Weismann's theory, however, as it implicates selection at the species level rather than the individual level, is not considered to be generally applicable", said John Maynard-Smith in 1976. Note that Maynard-Smith was a mathematician, not a biologist, and reduced the theory of evolution to simplistic models where only "fitness" and individual selection in toy systems determined the rules of the game.

The thinking behind this is that in order to limit adult lifespan, individuals would have to sacrifice their own continued lifespan and the continued production of their own progeny for the good of the group, and that the genes for such altruistic behavior towards non-directly related and actively competing individuals of the same species would be lost as individual competition for resources and progeny production would eliminate those with "altruistic genes" over time. This reasoning was taken as sufficient to eliminate group selection as a legitimate mechanism, and the game became how to limit lifespan without invoking group selection.

Clearly, an organism that lived longer would produce more offspring, so that lengthening the lifespan should be possible by individual selection alone. It is well-known that species have maximum lifespans such that a human may rarely live to 120, but will never (without some kind of intervention) live to 200; a blue whale may live to 200 years, but will not live to 1,000 years, etc. It is also widely known that within a species, various groups may have different lifespans, such that small dogs may live 12 years and large dogs less than nine years, while among dolphins the females live twice as long as males (more than 60 years). The fact that lifespan corresponds to the rate of predation in many animals is also well-known — mice, which have high death rates due to predation, have short lives and are sexually mature very, very young, while animals with low predation, such as squirrels, have very much longer lives and reproduce later. We will further discuss this, especially in regard to the naturalist Ricklefs' observation that "no single pattern of life, including development, maturity, and aging, varies among vertebrate species only by expansion or contraction of a common time scale."[28]

We will come back to David Neill's innovative ideas after we have discussed the origin of many of the modern "theories" of aging, but it is of interest that Neill tried to solve two problems with a single solution:

1. Why the ratio of age at sexual maturity is always a fraction of total lifespan, that fraction being specific for all members of a terrestrial vertebrate class (amphibian, reptile, mammal or birds), with the ratio of age at sexual maturity to lifespan (average or maximum lifespan, as they too are proportionate) increasing in that same order of the vertebrate classes, such that birds have the highest ratio of lifespan to age at sexual maturity (it would seem from ciliates that the same is obtained among them — a definite ratio between age, in cell divisions at sexual maturity and total clonal lifespan, also in cell divisions).

2. That evolution, at least as defined by natural variation and natural selection, i.e., Darwinian evolution, should have no mechanism for creating the ever-increasing complexity and the emergence of intelligence that we see evolving through evolutionary time. Evolution for adapting to an environment has a choice of more complexity or less complexity and there's no reason for more complexity — in fact it's the opposite.

As I said, we will discuss Dr. Neill's theory, which I believe makes some important points and explains some aspects of aging, but as Dr. Neill has bought into the idea that an aging program requires group selection and so is not generally applicable, this led him, as so many others, in the wrong direction.

The aging game

Sadly, to call the "study" of aging a game when an incredible amount has been learned of the mechanisms of cellular life is a tad unfair. However, if it were more than a game then it would have provided results, and it hasn't — at least not significant ones like an integral and not merely fractional increase in lifespan length. Yet, our group (Nugenics Research/Yuvan Research) has produced significant results, with the potential of conferring immortality, by not believing the "wisdom" that has accumulated from the time that August Weismann's posited "aging program" (in 1882) was rejected by mainstream science, based not on evidence, but on lack of imagination and perhaps a deep dread of a programmed death. That was really the display of incredible arrogance, as during the late 19th century the mainstream science knew barely anything of the basis of life. To me this is dismissing an argument because you cannot conceive that it could happen as suggested — proof by lack of imagination.

I believe this to be the reason the Church chose the reasonable hypothesis elaborated by Plato and then Aristotle, that the soul was non-physical — and as I said, almost all descriptions of the properties of Aristotle's "soul" describe the functioning of the nervous and endocrine systems — and the one part that Aristotle considered non-material, the ability to reason, we would equate with the "mind"; that the mind is "immortal" (though without feelings, only pure reason) is given no justification. Plato thought the immortal mind traveled back to the world of ideas to refresh itself and be reincarnated in another body (so you don't even own your soul). That some invisible something — your "thinking part" — was immortal, and it left the body to dwell in a Heaven God created as an eternal reward for being obedient on Earth or to suffer eternal punishment for disobedience, was not a belief that Jesus would have ever shared, but "reincarnation" even from the decomposed corpses of the dead was something that was beyond logic and comprehension, as though the universe was limited by humanity's ability to understand it.

The "soul" was an ancient (though ill-defined and severely limited) replica of a human being. Life in the underworld was, accordingly, a severely restricted form of life on Earth. The Greek philosophers retained this idea, but why? Is it that they too did not want to die? The ideas of eternal punishment and eternal bliss came from the Persian Zoroastrian religion. So really, the early Church seems to have been quite eclectic in choosing the best of mythologies to form a religion which gave the Church total control of a person, both during life and afterlife. If you listened and did as you were told (those were generally good recommendations — don't get me wrong, Jesus and even some "Church Fathers" had a lot of good things to say [and a lot of bad as well]), you might expect eternal bliss, but if you did the things the Biblical commandments and the Church's law told you not to do, assuming they were bad enough, the punishment was harsh (Dante, for example, classified the Inferno-worthy sins and sinners — the latter were usually his political enemies).

So, let's return to the reasoning. At first, the great geneticists and evolutionists Haldane, Hamilton and Fisher made the observation that older animals produced fewer progeny, and that fertility declined with age. Then during a lecture in 1951, Peter Medawar proposed that old age was the result of mutations accumulated in the germline that only acted in later, post-reproductive life, and so were resistant to selection. Now, though apparently unnoted, this explanation was circular reasoning: the proposal was that old age occurred because deleterious mutations occurring over evolutionary time were spared from being selected against because they only occurred at the end of life and were therefore resistant to selection pressures because reproduction slowed, but also assumed that old age and the post-reproductive period occurred because of the same mutations, the ones that cause reproduction to slow, that were spared by evolution as they only manifested at old age (post-reproductively). In this reasoning, the genes that caused aging resulted from aging, as the loss of reproductive potential with age is part of aging. Furthermore, how would this explain the fixed lifespans of animals (or plants — many fungi seem to be immortal, even large multicellular ones)?

In 1957, George Williams offered another explanation called "antagonistic pleiotropy", which might better explain aging and death.[29] In this case, William's "theory" relied on the then recently acquired knowledge that proteins can have several quite different functions for the same molecule; a molecule that is a metabolic enzyme in the cytoplasm can be a transcription factor (a molecule that controls the production of other molecules at the level of transcription [copying of DNA into RNA that will be translated into a protein]) in the nucleus. What if, hypothesized George Williams, during youth a protein displayed its "good" pleiotropic functions so that it increased the ability of the organism to reproduce successfully, but, in old age, that same protein now displayed its bad pleiotropic face to the cell, causing harm rather than good? According to Medawar's reasoning,[30] as the protein produced good effects during youth, it should be highly selected for, and such highly effective reproducers should dominate the next generations, even if their diminished abilities post-reproductively (due to the Mr. Hyde face of their pleiotropic proteins replacing the earlier beneficent Dr. Jekyll's) would cause aging and death.

This, of course, assumes that such proteins exist, which I stipulate they do not. However, furthermore, why and when does this Dr. Jekyll-like pleiotropic face suddenly "decides" to change its activities to Mr. Hyde's deleterious ones? At what age does this happen and why then? Vijg and Kennedy, in their paper *The Essence of Aging*, give what they describe as George William's own example of the workings of "antagonistic pleiotropy" thusly: "He hypothesized that a gene for rapid bone calcification during development would be selected in spite of the fact that this could also lead to depositions of calcium in the arterial walls, an age-related phenotype already occurring at middle age or earlier."[31] While it is true that the same gene may account for both phenomena — mineralization of bones during development ("good") and mineralization of veins during aging ("bad") — there is a basic misunderstanding of gene functioning involved.

While early geneticists sought the difference between humans and other mammalian species based on the differences between their genes, the truth is that, of the more than 20,000 genes in the human genome, a nearly identical number of nearly identical genes is possessed by every other mammal! It is not in most cases the presence or absence of genes in the genome that determines the phenotype (looks and behavior, etc.) of an animal, as it is the time, place, duration and extent of expression of these genes (together with the time, place, duration, etc. of other genes that might interact with them), among many other factors such as the presence of hormones or nearby cells that make a difference in phenotype. That is, it's not the presence or absence of the genes themselves, as they are nearly universally present in all mammals and largely interchangeable. In fact, even genes from simple organisms like *C. elegans* will adequately replace the homologous genes[*] of higher organisms like mammals and vice versa. So, the real question should be: why do veins activate the genes that result in their mineralization at middle and old age, and not before in youth (and inversely — and we know parts of the answer — why do bones stop their mineralization after a certain age)?

---

[*] "Homologous genes" are genes in different taxa (species, genus or even kingdoms and domains) that have a common ancestry and similar sequences of nucleotides, coding for proteins (in the case of mRNAs) with similar sequences of amino acids (or even simply similar 3D conformations). Homologous genes are usually discovered by searching for long stretches of sequences being identical or nearly, so in different organisms, these genes often have similar or identical functions.)

So, there is a difference in time and place (young bones, old veins) in the expression of the gene — the function (pleiotropic activity) doesn't change, the gene still recruits calcium to form deposits, but not at the same time and not in the same place during aging as during youth. So, according to Vijg and Kennedy, William's "iconic" example as it was just analyzed does not support William's thesis that the gene changed its behavior; the behavior remained the same, but it was just used for a different and destructive purpose during aging. The function of a hammer doesn't change whether you use it on a nail or on a human head, it still delivers a lot of kinetic energy to a small area.

While most researchers accept "antagonistic pleiotropy", it is more in the metaphorical sense. For example, the gene-product (protein) called TOR contributes to growth during development, but also prevents repair and maintenance during aging, though the function of TOR doesn't change — it still promotes growth processes and inhibits the transcription factor FOXO (which promotes the transcription of genes involved in repair and maintenance). The drug rapamycin, isolated from a fungus found on Easter Island (where those giant stone statues with long earlobes eternally guard the inhabitants), called by its inhabitants *Rapa Nui*, specifically inhibits TOR (which stands for "Target Of Rapamycin") and allows the expression of FOXO, therefore lengthening lifespan in several animal models (note that here again, the repair of existing damage increases lifespan).

The Russian scientist Blagosklonny believes the path to human longevity lies in inhibiting the mTORC1 complex (mammalian or "mechanistic" TOR complex), which is physically attached to the lysosome.[32] But we'll discuss how this is an unsatisfactory solution, as it can at best delay the inevitable. Note that while the "refusal" to allow maintenance and repair by mTOR (more specifically, the mTOR-C1 complex) might be considered an "out of whack" mechanism that unfortunately shortens lifespan, it can also be considered a perfectly functioning mechanism for increasing the mortality risk with age, an important part of the "aging program" dismissed by August Weismann's contemporaries.

Even before George Williams published his theory of "antagonistic pleiotropy", another mechanism through which aging might occur was championed by Denham Harman, an engineer so excited by the prospects of understanding aging that he took a medical degree to pursue it (his basic thesis was originally proposed by Dr. Gershman in 1954, but it was basically Weismann's observation that as organisms age they become less fit).

As an engineer, Harman knew that there were all kinds of destructive forces in our environment, like cosmic rays and normal background radiation, and that as time progressed, he would expect this accumulation of damage to reach toxic levels, causing aging. With the further additional knowledge by other researchers that the cell's oxidative metabolism in mitochondria itself produced toxic free-radicals such as the superoxide radical anion (O_2^-) and hydrogen peroxide (H_2O_2) (which white blood cells use to kill bacteria), providing a constant source for macromolecular oxidation damage, this idea was further supported. This was, in addition to "antagonistic pleiotropy" (as understood by most), the main focus of the anti-aging field: aging resulted from macromolecular damage (usually to DNA) due to random, potentially destructive, high-energy events (such as the transfer of a single electron to atmospheric oxygen) accumulating over time. So, somatic mutations (alterations of vital DNA sequences), were considered the mechanism of aging — but there was no definitive evidence of this occurring (though this might be a cancer mechanism).

Later, mitochondrial DNA damage displaced nuclear DNA damage as the suspected cause of aging, as the reduced mitochondrial efficiency seen in "aged" cells led to the increased intracellular concentration of reactive oxygen species (known as ROS; superoxide radical anion, peroxide, and various nitrogen compounds such as NO) and other highly energetic and hence destructive molecular species (we'll use ROS as a catch-all term for both ROS and those other highly energetic species). So, the concentration of ROS was highest in mitochondria, and the mitochondrial DNA-repair systems were inferior to nuclear systems.

In fact, the SENS Research Foundation (SRF) — an institution heavily involved in aging research — has shifted genes from the mitochondrion to the nucleus (as throughout evolution, mitochondrial genes have shifted to the nucleus without external help) in an effort to protect them from ROS damage. Clearly an interesting experiment, but not likely to have anything to do with aging. It certainly won't help those of us now living, but might (though I doubt it) lengthen the lives of our children if such tampering with human evolution, and with totally unknown and unknowable results, might ever be permitted (I hope not, though I wouldn't object to seeing an entire animal [say a mouse] whose cells have been so reconfigured).

So, in the preceding paragraphs I mentioned DNA repair, and indeed, as random damage was the presumed mechanism of aging, the introduction of the cell's ability to repair such damages enzymatically would seem to have a chilling effect on all stochastic theories of aging that assumed aging and death resulted from precisely such causes as accumulated damage: if the cell were able to repair macromolecular damage (and the only really essential damage was to DNA, as any other part damaged could be replaced from instructions carried in DNA), then what caused the degenerating conditions in aged cells?

One popular answer was that the repair abilities of the cell were not up to the task of repairing every kind of damage, or repairing it correctly. Thus, lifespan should depend on how efficiently DNA damage is repaired; there were arguments, back in the '80s, that a mouse did not live as long as a human because its repair endonuclease was not as efficient as the human variety. A "repair endonuclease" is an enzyme that recognizes DNA damage and puts a "nick" near it (a "nick" is a single cleavage of a single strand of double-stranded DNA), so that other enzymes can recognize, remove and replace it with correctly sequenced and undamaged DNA (the repair DNA polymerase enzyme — the enzyme responsible for performing this repair — gets its information from the complementary DNA strand). But this description of DNA repair is more applicable to the much simpler bacterial systems, as the DNA repair process is more complex in eukaryotes, with DNA surrounded by histone and non-histone proteins as well as RNA strands of various types.

In these early evolutionary theories of aging, stochastic damage represented the mechanism of aging. The environment contained enough sources of disruptive, high-energy radiation and molecules to eventually damage vital biomolecules. However, with time it became clear that Nature was not unaware of this constant attack on cellular integrity and developed means to counteract them — DNA repair, other forms of repair (the chaperone proteins that reform misfolded proteins, for example) and a host of enzymes to govern the redox potentials of the various cellular compartments (i.e., nucleus, cytosol [not including organelles of the cytoplasm], mitochondria, endoplasmic reticulum, etc.).

In fact, DNA repair was so fundamental to life that even those strange creatures, the viruses, which hang out on the border between living and non-living, even those viruses that attacked bacteria (and probably existed before the first eukaryotic cells), had their own DNA repair systems, and of different types for different types of damage. The cell was not a passive "Nature's punching bag", but could respond to damage with repair.

Thomas Kirkwood recognized that the existence of DNA repair and other sorts of cellular repair changed the argument. How could random, accumulated damage be the cause of aging, if such damage could be repaired by the cell? Clearly (as was first pointed out by August Weismann), the soma (body) and germline cells separated early in development, and while the germline cells were capable of the nearly perfect repair that allowed their contents to remain the same for millennia (notable mutations being the exceptions rather than the rule), somatic cells had limited lifetimes, and according to the prevailing stochastic theories of aging, limited ability to perfectly repair themselves, that inability resulting in aging and death. Recent evidence shows that this is not true, that germ cells age and the age of the earliest embryos are like the ages of their mothers. Thus, the germline does age, but when gastrulation (an early embryonic stage) occurs, and the embryo's own genes control its development, the cells of the embryo reset their ages to zero (and apparently carry out perfect repair).[33]

At that time (around the 1980s), it was clear to most students of evolution that the immortality of the individual organism was not one of its goals, but the immortality of the species was. But why not make the individual as immortal as possible (at least repair defects in all of its cells, somatic and germline)? The answer Kirkwood gave to that was that in Nature, the energy available to any organism was scarce and had to be used only for the most essential functions, and as the history of life is more than 3.5 billion years old, it is clear that the single most important function was to reproduce to ensure the continuation of its kind.

So, the argument for Kirkwood's "theory" known as *disposable soma* is that as Nature provides only scant energy on average to any organism in the wild, that energy is allocated to reproduction, allowing only the germline cells, segregated from the "soma" (body) during embryonic development, to use the scarce energy available to effect error-free repair as the germline must last the length of the species, while energy lack severely limited the apportionment of repair capacity to the somatic cell, as the individual body was peripheral to the survival of the species. It needed to live long enough to reproduce, and in mammals and birds, to raise their offspring to self-sufficiency (the female octopuses aerating and protecting her brood is comparable).

So this *disposable soma* "theory" (in fact Kirkwood doesn't pretend it to be a theory in the scientific definition, in that it is not the results of several proven hypotheses that converge on a greater understanding, nor in the sense that it is predictive) simply provides an "explanation" of how an organism that has the cellular ability, as shown in its germline cells, to perfectly repair itself, fails to do so and thus ages and dies.

These "theories" presented above are the so-called "evolutionary theories of aging". The theories of Medawar posit deleteriously genes only acting at the end of life (through a faulty process of circular reasoning that essentially says that old age causes old age). And yet, there are such genes; for example, the thyroid-hormone carrier, transthyretin, is overproduced in aging, though it no longer has the function of transporting thyroid hormones, and as it is an amyloid-forming protein it contributes to the amyloidosis seen in aging; however, this is not a mutation in the transthyretin gene, but in its regulation. We do not have proof of mutations that suddenly onset at middle or old age, we simply have different age phenotypes at different ages. If Medawar's understanding of aging were correct, and we were to search for a path to immortality, then it would behoove us to find and correct those harmful mutations that only appear at the end of life, but there are no such mutations — no immortality to be found here.

If Williams' theory were true, there would be absolutely no way to control the Jekyll and Hyde character of genes that are beneficial in youth and deadly in later years. In this model, there is no clue as to when a gene would change its character to some deleterious form. We have explored the late George William's example of an iconic gene showing "antagonistic pleiotropy", the calcium-binding protein that mineralized bones in youth and veins in old age, but found that far from being an example of one gene assuming different forms and functions, it was rather one gene used in different ways in different tissues at different stages of the lifespan. The gene that originally mineralized bones by calcium binding at one stage of development in bone tissue was simply used for a different purpose in a different tissue (vein) at a different post-adult developmental stage. Whether the purpose of that change in usage was benign or baneful is not decided by that calcium-binding protein whose activities remain what they were (calcium binding), but by the cells that produced them.

One might well ask (and we'll have an answer) why venous cells should want deposits of calcium, but should not ask the gene, as it had no part in that decision, but rather the cell that invoked it. A road to immortality, or at least substantial life extension, could not result were this model true. I personally don't believe there is any truth in it, and the argument could be better expressed by *genes may change their functions in a multidimensional space that includes tissue type, developmental stage, organ, and intercellular and intracellular environment.*

Finally, the most comprehensive of the stochastic "evolutionary" theories of aging (a strange sort of evolutionary theory, without biology or the history of life or even group selection), which allowed mathematicians like Thomas Kirkwood to engage in the evolutionary parlor game, is the following: if we could convince cells that they'll always have enough energy (we're not living in the wild, and lack of energy — food calories — is the least of most industrialized nation's people's worries), might they start using that energy to repair cells? There's no clear way to do that, and it would seem the cells would remain content with the immortality of the germline (or part of it anyway). But there is some possibility of immortality here (merely change the conditions to allow perfect repair in cells other than embryonic cells), by simply changing the timing of when somatic cells lose repair capacity, and basically that is how E5 works (the compound we used to rejuvenate rats) — by fooling the cell into thinking it's the cell of a young animal.

What separates these widely believed theories from another that I will name, after a brief interlude into the natural world, is that in all these evolutionary "theories", lifespan has no relationship to the animal in its environment. Lifespan length is, in Medawar's estimation, determined by random mutations the species' experienced in its long history, mutations protected from selection by being at the end of life, that cause the end of life. So, here, if we suspend disbelief, we may assume that a species' lifespan is the entirely random result of deleterious mutations the species acquired in its history that express themselves during life-stages we recognize as the species' "old age". There is no relation between lifespan and niche — lifespan is randomly derived.

In the extended version of George William's "antagonistic pleiotropy", the original contention of a pleiotropic protein is replaced with a pleiotropic "system", like the yin-yang, an eternal battle between growth, as represented by the mTORC1 complex, and maintenance and repair, as represented by the FOXO transcription factor (which controls the elaboration of many such repair and maintenance genes). Yes, in youth, the mTORC1 complex headed the cell in the right direction, growth, but after growth stopped, the mutual inhibition of mTORC1 complex by FOXO, and FOXO by mTORC1 plus the continued activation of mTORC1 prevents cellular repair when needed. This is to many a vindication of "antagonistic pleiotropy", but actually it is not, as the function of the mTORC1 complex remains to encourage growth and repress repair, even if that's not the best solution (from our viewpoint) for the cell or the body.

So, it's perhaps not surprising that Dr. Blagosklonny believes that turning down mTOR (which is directly targeted by rapamycin — and rapalogues — in the mTORC1 but not in the mTORC2 complex) should increase lifespan, which it does, but only to a fractional extent and with side-effects. The real problem, however, is that because he misunderstands aging, Blagosklonny doesn't realize that suppressing the mTORC1 complex has only marginal effects in higher organisms. Again, the theory of "antagonistic pleiotropy" also randomly assigns lifespan to an unknown mechanism that causes a protein, or a gene regulatory network (GRN), to change to a deleterious form at some particular stage of an organism's lifespan without a clue as to why or when that would happen.

And finally, by cutting off repair capability to somatic cells, the "disposable soma" limits the lifespan to the probability of cells being inactivated or mutated with time and no mechanism is given for determining lifespan other than pure chance.

The knowledge that there are species-maximum lifespans proves that lifespan is not determined by pure chance. In my view, a cartoon-like representation of life as presented by a group of neo-Darwinian "evolutionists" who want to cram the whole of evolutionary science into a tight little box held together entirely by "natural variation" and "natural selection at the level of the individual", and read the results of cartoonish life-processes from their simplistic computer models, is not the correct way to base theories.

Regarding the problem of extending life in the "disposable soma" model, if the repair mechanisms that supposedly give perfect fidelity to the germline (actually proven false as noted above — it is the early embryo which resets its age to zero) cannot be invoked by the soma, the problem can then be solved by preventing the creation of damage. This led (in part, together with Harman's ideas of aging damage accumulation) to the emergence of the idea that compounds that prevent the formation of (or intercept and disable the damage-causing) free radicals and other ROS would slow down the aging process. Antioxidant vitamins and even enzymes like superoxide dismutase (which would simply be digested like any other protein) were taken in large quantities in order to slow aging — but there were no apparent effects (though some effects were obtained in simple organisms and in cell culture), no increase in lifespan nor (as many assumed) decrease in cancer rates. In fact, experiments assessing the effects of the antioxidant beta-carotene on cigarette smokers (which should have been a no-brainer according to the prevalent theories of damage accumulation, causing aging and cancer) was discontinued due to the excessively high rate of lung cancers in those who were given doses of the antioxidant beta-carotene as compared to those given a placebo ("sugar pill").[34]

The various compounds that were antioxidants, free-radical "traps" designed to neutralize ROS — many of which showed anti-aging effects in cell culture and in simple organisms — showed no or limited effects in people, such that Scientific American, a popular science magazine, devoted an issue to warning readers of the lack of any proven results from the use of these "nutraceuticals" and other over-the-counter (OTC) vitamin and mineral combinations together with antioxidants that were flooding the market as the baby boomer generation started to age. There were many excuses given at the time as to why these antioxidants didn't work — some advanced the theory that the artificial antioxidants turned off the production of natural ones. But clearly the road to immortality, or even significant life extension, did not lie in that direction. If ROS were really the culprit in aging, then why didn't all those antioxidants work? Animals so treated must have had their cells in near perfect condition with all that protection, right? One hint is that they worked well enough in cell culture, but not in animals. It might be that cellular and organismic aging work by different mechanisms. The short answer, to which our own research applies is yes, but not exactly, as there is feedback between the cells and the organism.

What does life have to do with it?

All of the above "evolutionary" theories of life are, as I said, the result of mathematical models abstract from the realities and complexities of life and purely accidental events — such as the several mass extinctions in which substantial fractions of living things died, entirely changing the environment and permitting new regimes of life to replace the old. For example, the roaring, giant dinosaurs died out while the tiny, shrew-like mammals that ate their eggs became us, following the twin catastrophes of the asteroid crash in Mexico's Yucatan Peninsula and the tremendous outpouring of lava and gas in the Deccan Traps region of India. It would be difficult to include these events and predict their outcomes in simple digital representations of the living world, like comparing Donald Duck to a real duck.

One article I remember reading showed that a shortened lifespan could lead to a greater average population density (more biomass) if animals were geographically distributed. I was not so amazed by the result (though it was the first demonstration by computer modeling of the fact that shortening lifespan could be advantageous to the species — that alone was a breakthrough, it came out of MIT), but I was amazed that prior computer models that researchers used to validate "theories of aging" (aging "hypotheses", more correctly) did not include geographic distribution. Does that mean these hypothetical animals developed in a one-dimensional world? How could anyone accept such a model of "living things" in competition for "resources" in a one-dimensional "world" as sufficient evidence to base such an important topic as lifespan on? And yet people did.

Before I leave the topic of "evolutionary" theories of aging, it's important to mention that the suggestion that group selection is not an important evolutionary force is contradicted by a phenomenon that is widely observed in the present world but not widely known or noted during the periods when classical Darwinism first arose, which is the introduction of alien species. It is my contention that when an Asian snakehead fish invades a USA pond, not even the meanest sunfish or toughest perch will stand a chance against it. In the eastern USA, the English sparrow has nearly totally replaced the eastern bluebird, and now, the Burmese python is becoming a top predator in the Florida Everglades, competing successfully against the alligator, and we don't know what will become of this land as sea levels rise.

So, my point is: natural events (or human-made events, nowadays), including the emergence of new species and catastrophes (such as asteroid crashes), made group selection an important source in the change of populations in regions over time. Group selection may lie at the heart of species changes. When oxygen-producing photosynthesizers started "poisoning" the atmosphere, there was the first great die-off of anaerobes, and those few that remained were now confined to hidden places on the planet, and the aerobes now had a new and more powerful source of energy (oxidation with oxygen) if they could learn to use it, and we did.

So, again, if you want to study aging, do you want to start with toy mathematical theories that change predictions based on the ingenuity and painstakingness of the modeler, and not actually look at the aging of biologically aging organisms? If we do that, we notice that there is a wide range of lifespan distributions; some protozoa and fungi are immortal, with the mortality of ciliate organisms quite restricted (as we noted). Also, there are sponges — animals having about ten different cell types and which might be labeled as colonies of choanocytes (unicellular, flagellated "collar" cells) in which amoebocytes digest and distribute food (both choanocytes and amoebocytes can become gametes and give rise to all cell types) — who can live more than 10,000 years, and the more complex flatworm has that potential as well. There are cold water clams that live 500 years, and lobsters that live a couple of hundred years (are they non-aging?). The female Greenland shark doesn't become fertile until she's 150 years old, which also indicates a very long lifespan.

However, by the time we get to land-dwelling vertebrates like ourselves, lifespans shorten considerably. But what determines lifespan? The evolutionists tell us it's unavoidable though random, but that is not what naturalists observe. There are patterns in lifespans that are predictive and explanatory — patterns that connect lifespans to ecological roles; yet that cannot exist if the lifespan is not open to selection, if it is determined by purely random processes. So, let us discuss two theories that explain lifespan length as a non-stochastic process, the "rate of living theory", which explains much, but has multiple exceptions, and more of Dr. Neill's theory, which also explains a lot, but as our evidence indicates, is pointed in the wrong direction. We will see how a few slight changes in perspective lead to what is increasingly being seen as the road to immortality.

Okay, I know you are tired of these theories that decide what life can do restricted by "evolutionary thinking", but I want to delve into these two theories. Not because they have justification from the viewpoint of evolution, but because they present an entirely different interpretation of lifespan, one based on the observation of aging in organisms. And finally, after these two theories, I will present my own.

The rate of living theory

Nearly everyone is familiar with this oftentimes popular theory of aging. It goes, in general, something like this: animals all live a fixed number of heartbeats, or breaths. It is apparent that small mammals have short lives, with exceptions; they mature (sexually) quickly and die young. On the opposite extreme, large mammals such as blue whales live much longer and mature later. The scientific basis of the rate of living theory began with the observation by Max Rubner in 1908 that larger animals lived longer lives, and had a slower metabolism. Later, a relationship between the mass of an animal and its metabolism was proposed. This was called the Max Kleibers' Law; it said that the basal metabolic rate (BMR) of an organism is proportional to its weight to the 3/4 power (which means cubing the fourth root — but it's close to 1). So, we could now relate metabolic rate to weight (mass), and so lifespan to metabolic rate. The basic rate of living theory hypothesized that there is an inverse (negative) relationship between lifespan and energy expenditure as shown in Figure 11.

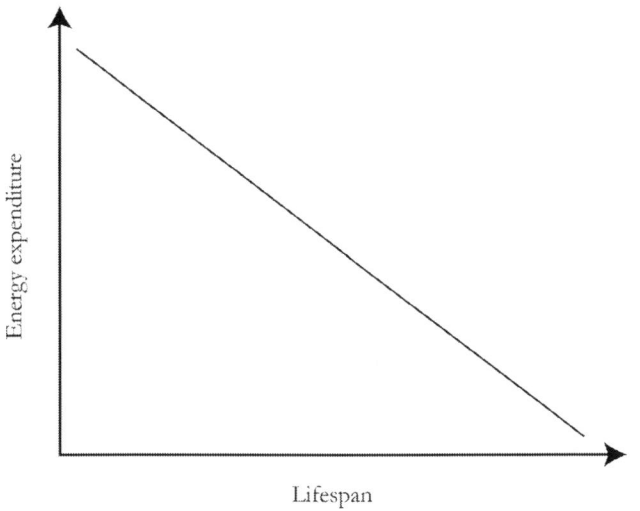

Figure 11: Representation of the rate of living theory basic idea.

One added attraction of this theory came with the knowledge that the generation of energy through mitochondrial oxidative phosphorylation has the side-effect of 0.1% to 3% of the oxygen molecules used in the process only passing one electron to oxygen — forming the superoxide anion free-radical. In chemical terms, this means that (at least) one atom of the molecule possesses an "unpaired" electron (electrons have a strong preference to be paired with another electron of opposite spin) — looking like this: $O_2^{\cdot -}$, where the superscript dot represents the unpaired electron and the superscript dash its single negative charge. So, taken together, we have the picture of a "faster living" organism producing more ROS and hence more damage, resulting in more rapid aging and a shorter lifespan, as in some organisms 3% of the oxygen molecules in the ATP synthesis process form ROS as opposed to "slower living" organisms in which this percentage is only 0.1%. And among invertebrates there were several experiments that seemed to agree with this conclusion.

Raymond Pearl, in his 1928 book *The Rate of Living*, discusses measurements agreeing with and amplifying the ideas of Rubner using domestic animals as an example. Of course, the problem here is that correlation isn't causation, and that while the correlation between size and longevity may validly reflex some degree of causality, the inference that lifespan and basal metabolic rate are more than correlation is not supported by the data; when size is taken into account, there is no relationship between basal metabolic rate and lifespan.[35] Of course, whether the basal metabolic rate is a good measure of lifetime energy expenditure is a question. One of the more interesting early experiments showed that taking the wings off of flies, hence severely limiting their energy needs (as flight takes considerable energy), significantly increased their lifespan (how many school boys have done this for the wrong reasons?). In the roundworm *Caenorhabditis elegans* (a "workhorse" of aging science), Cynthia Kenyon found that mutations of single genes can significantly extend the lifespan leading to the thought (at least in my head) that there are genes that functioned to limit lifespan.[36]

One of the early objections to the rate of living theory came from comparing small mammals to birds, wherein the birds far exceed the metabolic output of the small mammals, yet have longer lives. This was later explained as the increased efficiency of bird as compared to mammalian mitochondria.[37] Perhaps the same applies to bats, which are unusually long-lived considering the high metabolism required for flight. But there is a basic problem in comparing different classes: the same capabilities for repair are not present to the same extent in different classes of vertebrates (for example), so in order to assess whether or not one particular variable determines lifespan, it would be necessary to isolate that variable.

What the naturalists say

The naturalist Robert Ricklefs tested certain assumptions of aging by using a statistical aging model which better fitted the data from animals in the field. He found that animals with a lower initial mortality rate had a slower rate of aging, but that a greater proportion of deaths in these species results from old age, which means that there were still potential life-extending factors that were selectable. This ruled out mutation accumulation and antagonistic pleiotropy as causes of aging, and led Ricklefs to conclude that aging results from "wear and tear", on the one hand, and genetically controlled mechanisms of prevention and repair, on the other. "Evidently, remedies for extreme physiological deterioration in old age either are not within the range of genetic variation or are too costly to be favored by selection.",[38] was his conclusion — much in line with disposable soma. Here, though Ricklefs knows that both time to sexual maturity and lifespan depend on predation, he can find no connecting mechanism for them. If David Neill is correct, the connecting mechanism is the frequency of a "life's timer". This statement makes the assumption that "remedies" are needed for the physiological deterioration due to aging, though Nature seems to disagree, as it can be seen in the progressive down-regulation of repair and continued upregulation of destructive cytokines and chemokines (like inflammatory cytokines, and the chemokine eotaxin).

However, Ricklefs, and later João Pedro de Magalhães (Dr. de Magalhães has researched nearly every aspect of aging and lifespan), reached the same conclusion; the primary determinant of lifespan length (does de Magalhães believe that lifespan is a selectable trait?) is predation; hence the reason that small animals have short lifespans is that they are more susceptible to predation (small animals not subject to predation, like the naked mole-rat, have long lifespans).

However, de Magalhães added something more, when he confirmed many studies using much larger databases than in the past, concluding that the average or the maximum (as they are proportional) lifespan is a function of "developmental time". De Magalhães wrote after a careful survey of hundreds of mammalian and bird species that "Overall, these results indicate that, independently of body size, developmental time is strongly associated with maximum adult life span."[39] In fact, de Magalhães, as have many other authors, established functions whose input is time until sexual maturity and whose output is lifespan, for both mammals and birds.

So, let's stop a second to understand this new observation as it was derived from data and not theory; by "developmental time" is meant the time from fertilization to birth added to the time from birth to sexual maturity. The evidence (which has a long history) that lifespan post-sexual maturity is a function of the time period from fertilization to sexual maturity in both mammals and birds is overwhelming with thousands of species included. Why might this be? Let's go back to what David Neill thinks about this.

One further remark that Drs. Ricklefs and Wikelski make is worth reading in its entirety: "The rate of reproduction, age at maturity and longevity vary widely among species. Most of this life-history variation falls on a slow-fast continuum, with low reproductive rate, slow development and long life span at one end, and the opposite traits at the other end. The absence of alternative combinations of these variables implies constraint on the diversification of life histories, but the nature of this constraint remains elusive."[40]

So, again let's unpack this statement. They are saying that in spite of the wide ranges in age at maturity, reproductive rates and longevity of animals (they chose birds because of the abundance of data), these traits don't independently assort but are grouped together, at both ends of what they call the fast-slow spectrum — they either live short lives, mature early and reproduce rapidly, or they have long lives, with late maturation and low rates of reproduction (think mouse vs. elephant). Thus, we don't have short-lived animals with low reproductive rates and late maturity, or long-lived animals with early maturity and high reproduction. Put another way, this is a way of saying that there exists a proportion between these rates (maturation, immature period and lifespan) that has no clear explanation.

David Neill, part 2

It was David Neill's stated aim to explain the fixed ratio (or at least the functional dependence) of post-mature lifespan on developmental period as described above. The way this was solved was quite different from the assumptions of mutation accumulation or antagonistic pleiotropy, and can best be explained by looking at the illustration of the process in Figure 12.

The three blank "phases" at the end of the lifespan represented in Figure 12 is to represent the fact that the life-phase producing oscillations are themselves not restricted, but that their cessation results from the death of the organism from a lack of further developmental stages. So, the assumptions made here are that the posited "Life's Timer" divides life into equal-length phases, and that the number of pre-maturation phases (four in the diagram) to the maximum number of post-maturational phases (nine in the diagram) for each class of vertebrates remains fixed (though exceptions exist).

It is further assumed that the classical "accumulation of damage", i.e., "wear and tear", is the major factor in death — as "during the PM (post-mature) period there is a graded de-investment in cellular maintenance and repair" with accumulating damage (at the cellular level) ultimately leading to death.[26] At later periods of adult development, the disinvestment in repair and maintenance should grow and hence the accumulation of damage should accelerate with age. This has been shown many times (as for example, the resistance to radiation damage declines with age), but the result of that — a growth in mortality rate with time — increases exponentially.

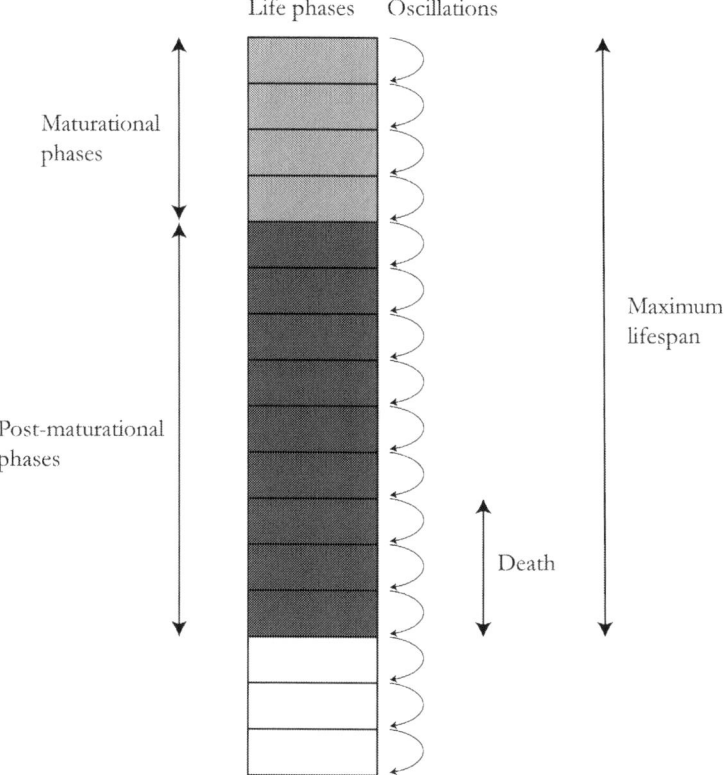

Figure 12: Representation of life phases driven by oscillations leading to death. Redrawn.[26]

Neill proposes this based on his own 2010 paper which he sums up as: "Rather than a genetic pathway/program, this theory proposes a longevity timekeeper that can delay the rate of accumulative cellular damage during the post-maturational period of life."[26] So, here, a "Life's timer" (which according to Neill must be intracellular and present in all cells) sets the pace of life. Several oscillatory circuits that may work as clocks are mentioned, many being ultradian clocks whose oscillations occur several times a day, and Neill also mentions the circadian clock, which oscillates every 24 hours (or thereabout).

In the diagram in Figure 12, Neill stipulates that each oscillation defines a "phase", but what does phase mean in this sense? Clearly the immature phases correspond in some manner to the developmental phases — zygote, embryo, fetus (and there are many stages without names). However, assuming that the timer is the circadian clock — which I will give evidence for — there are no 24-hour life phases that we know of (except the cell cycle which is regulated by the circadian clock, and its effect on the redox state of the cytosol and nucleus). And what of the post-mature phases? Do we know of any post-mature life stages? I will address this more fully later, but the short answer is yes — young adult, middle-aged and old are some rudely delineated post-maturational life phases.

In David Neill's model, there are thousands or tens of thousands of phases, but what do they mean? To Neill, the "Life's timer" (LT) controls the development in immature stages and the deterioration in post-mature organisms. As we will see, the LT controls developmental progression — the coordinated aging of organs simply being age-phenotypes that are consistent (with individual variation) across a species, leading to death.

Aging researchers and development biologists differentiate between development — where, in some cases, but not all, an animal develops into a more complex and capable form — and "aging", where the body assumes a less complex and less capable form. However, this is a distinction without a difference; in many species, complex motile forms become simpler sessile forms, filter-feeders that actually eat their own brains to save energy, but we still consider that part of the normal development of such species. To assume that aging is simply the deterioration of the young adult neglects many age-specific traits that are not deleterious (for example, the different life-tasks required by animals in middle age as compared to young adults, at least in higher vertebrates), and for the species, there are enormous amounts of evidence that lifespan length is a species trait that is conserved and selectable. For example, in the African turquoise killifish, their lifespan is closely adjusted to the length of time the pools created during the rainy season remain liquid, the difference in different parts of Africa varying by months, yet the killifish have life cycles — from birth to reproduction to death of old age — that ensure the fish gets to live full lives, reproduce and die, before their ponds turn to hard mud.

The passage through a young adult stage, when an organism proves itself capable of securing a mate (reproduction), middle age, where in the case of many mammals the reproducing middle-ager has the responsibility of providing for mate and offspring, and finally old age, where the organism can no longer compete against the young and hands-off leadership to the next in line (though unwillingly and as the result of a battle in many cases), is the common pattern in mammalian lifespans, much as they are in the lifespans of invertebrates. The question is whether this pattern is programmed and whether it can be changed.

Big data and bigger ideas

The "law of large numbers" is the idea that when the numbers of observations grow large enough, the intrinsic probabilities, rather than the mere statistically implied results of small experiments, will be revealed. When using small numbers of animals (*C. elegans*), we find that 50% of the population dies within two weeks; when we use large numbers (more than 100,000 animals), the "law of large numbers" makes that same result into the probability of dying within two weeks. Thus, it was of particular interest that Walter Fontana's lab at Harvard (Fontana Lab), with the assistance of a mathematical biologist, Nicholas Stroustrup, discovered the mathematical basis of age-related decline.

The thesis was simple: Fontana at Harvard and his collaborator Stroustrup developed an automated system that could detect the death of each worm of more than 100,000 *C. elegans* kept in various different environments to within 20 minutes of their death. This led to mortality curves of unprecedented accuracy. In the introduction to their paper, they say: "Here, by collecting high-precision mortality statistics from large populations, we observe that interventions as diverse as changes in diet, temperature, exposure to oxidative stress, and disruption of genes including the heat shock factor hsf-1, the hypoxia-inducible factor hif-1, and the insulin/IGF-1 pathway components daf-2, age-1, and daf-16 all alter lifespan distributions by an apparent stretching or shrinking of time."[41]

Personal note — when I tried to promote my own idea of this by talking about a clock (much in the manner of Neill), Fontana objected to having my comment about his treatments speeding or slowing an "aging clock" being published in Nature and rejected the idea of a clock or timer. A second set of editors agreed with me, but a third said that my comment, while relevant, was too important to remain a mere comment, and suggested that I developed my ideas as an article (well, that's happened and is happening as you read). What I really thought was that they just wanted to shut up an unknown in favor of a Harvard professor — I'm not sure I blame them.

The evidence for this shrinking or stretching of time (biological time) is apparent and striking, and it's based on what is known as "survival curves". These curves simply display the percentage of the original population (plotted on the vertical axis with the population starting at 100% and ending in 0%) remaining after a period of time starting from hatching (that time is plotted on the horizontal axis, as seen in Figure 13). If we look at the top graph, we see the normal mortality curve (that is, the control group of worms raised under standard conditions) in the solid line; the population starts at 100% and slope downwards, in a manner resembling exponential decay until it reaches zero. It should be recalled that the huge number of animals that make this study so significant that they change experimental results into hard probabilities of future survival due to the "law of large numbers", no longer applies at the tail of the curve, where the numbers of surviving animals are massively reduced.

In the two other curves in the top graph, the worms are placed in environments or have mutations that hasten aging, i.e., shorten life (various ROS emitters, warmer temperatures, disruption of heat-shock genes, etc.), or are raised under conditions that lengthen life (cooler temperatures, DAF-2 mutations and reducing agents like N-acetyl cysteine in their environments). In the bottom graph of the illustration, the Greek letter λ is equal to the lifespan length of the experimental group divided by the lifespan of the control group, so assuming the lengthened life (dashed curve in the illustration) to be twice the control lifespan, $\lambda = 2/1 = 2$. Now we can create the function $r(t) = t/\lambda$. If then, instead of graphing each survival as the percent survival versus the time from birth, we divide that time (t) from hatching by λ, forming $r(t)$, we see that all survival curves formed in this manner overlap!

As large numbers of animals were used, we can take any point on these curves, and define an age-specific mortality $d(\log t)/dt$ (accurate to within 20 minutes) that we can define as a "phase", and the inflection point when mortality begins to increase might be one such phase; it appears to occur at about the first 1/3 of the total lifespan in the control group — and can so be seen to occur in the first third of total lifespan under all the different conditions elaborated, whether shortened by heat or lengthened by mutation. When percent survival is plotted against $r(t)$ (rather than "t"), all survival curves superimpose on one another as shown on the bottom graph of the illustration. It could be concluded by this that no matter what the ostensible cause of death, the difference from the untreated control is just the length of the life phases, as all survival curves have an identical shape when plotted against r(t). The only factor that affected the survival of these worms was the proportion of their lifespans they passed through.

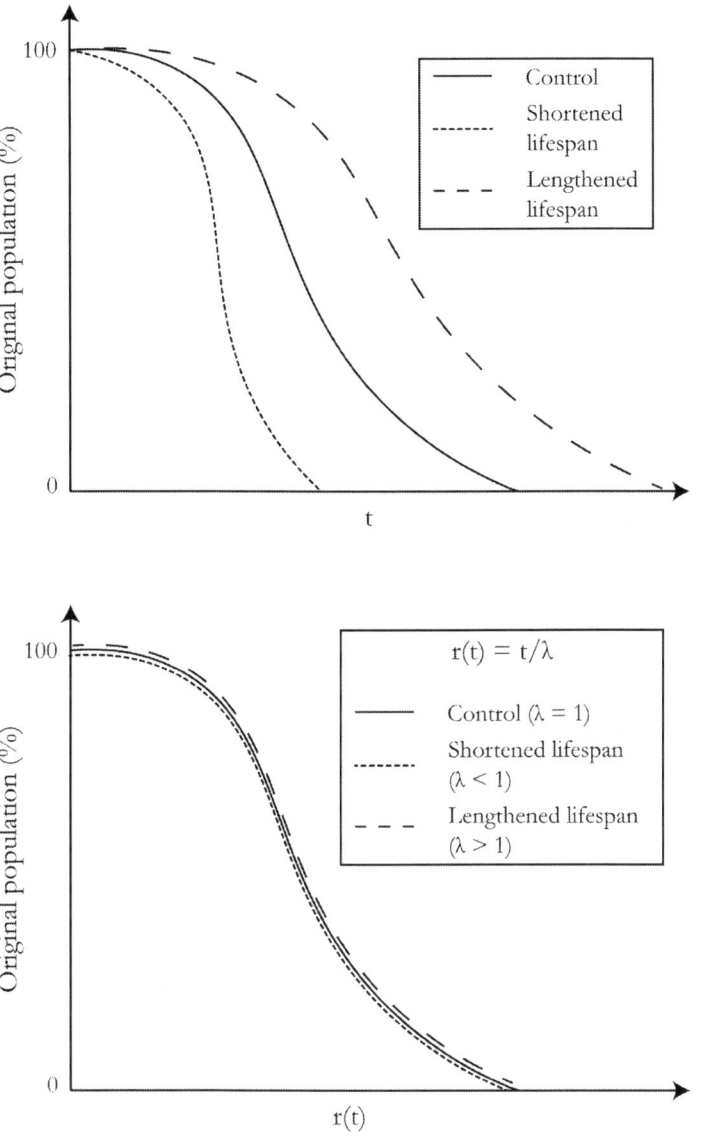

Figure 13: Survival curves of control, shortened and lengthened lifespan for three populations of *C. elegans*. Redrawn.[39]

Thus, one single factor determined mortality from all causes. What Stroustrup and Fontana's study revealed is that lifespan is a function of "resilience"; while no further mechanism for "resilience" is revealed by the study, it concluded that all-cause death results from a single variable associated with lifespan as compared to lifespan under more favorable conditions.[39] Stroustrup and Fontana call that variable "resilience" and hold its loss responsible for all causes of death. Resilience declines with age, until it is not sufficient to save the life of a challenged cell.

So, is "resilience" merely a word, an abstract concept for the loss of something defined as the ability to survive challenges? What then is responsible for "resilience" — what is it really? In David Neill's estimation (his theory of aging), would it be the continued ability of the LT to extend youthfulness, or the loss of an ability to overcome adversity? And still what is this mysterious LT? Also, does the same apply to higher organisms? There is some evidence that I will discuss later. If instead of looking at survival curves we look at mortality curves the result is represented by the graph of Figure 14.

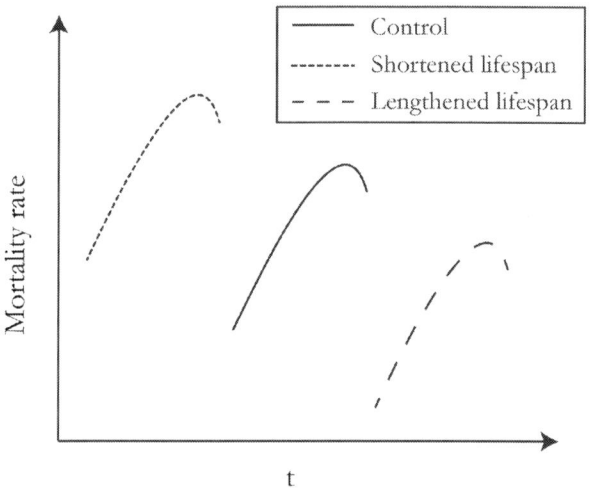

Figure 14: Mortality rate as a function of time for three populations of *C. elegans*. Redrawn.[39]

We see the similarity in these curves, though in reality the lifespans range through more than an order of magnitude[39] — this is especially striking as this is a semi-log plot; the apparent linear nature of the lines reveals an exponential increase in the hazard rate throughout their lives. The downward "hook" that occurs at the upper end of each curve may represent a diminution in the force of aging, or since the majority of the large populations are dead at this point, it may simply represent the persistence of a naturally resilient variant population.

6

Why do aging cells die?

In all aging theories discussed this far, aging results in the loss of cells, and though we haven't discussed it yet, it results particularly in the loss of "stem" and "progenitor" cells — cells designated to produce the somatic cells needed for continued life. In some cases, cells behave abnormally, either by attrition of their telomeres past a "critical length", or by excessively producing oncogenes (genes that result in cancer when expressed by some cells — most are growth factors), or in other ways, cycling cells become "senescent cells", which are taken out of the cell cycle.* These now non-reproducing and apparent functionless senescent cells, innocent-seeming, but doddering old cells, taken out of the cell cycle and put into retirement, are not really dead or dying at all, but are metabolically very active secreting products (the phenomenon of senescent cells secreting potentially harm-

---

* The cell cycle, or cell-division cycle, is the series of events that take place in a cell that cause it to divide into two daughter cells.

ful products is called SASP — senescence-associated secretory phenotype). Those products promote other nearby cells to become senescent, and these senescent cells also secrete enzymes that disrupt the cell matrix (which binds cells together). This, in turn, allows the migration of epithelial cells that have transformed; the EMT (epithelial-mesenchymal transition) gives epithelial cells (cancers based on epithelial cells are called *carcinomas*, the most common sort of cancer in humans) the ability to leave their locations (normal epithelial cells die when removed from the single-cell thick sheets they form) and migrate to the blood stream or the lymphatics (a system parallel to the circulatory system that provides for cellular waste removal and produces immune cells to fight invaders). Once they left their place of origin, these transformed epithelial cells (they now look like mesenchymal cells, through the EMT) are able to establish micro-colonies throughout the body. If these microcolonies survive, they become cancerous tumors.

The transformed cells of cancer have an advantage other cells don't have: they are not programmed by the body (normal cells are obedient to the body's rules) and they are constantly prodded (often by themselves) to reproduce. Normally, cells that are badly damaged will self-destruct by a number of intrinsic means — apoptosis is the best known, ferroptosis is also well known, as is simple necrosis, but cancer cells don't do that.

In 2013, López-Otín et al. wrote a summary, called *The Hallmarks of Aging*,[42] of what is believed, by the overwhelming majority of researchers, to be an exhaustive review of the literature. Like its similarly named predecessor, *The Hallmarks of Cancer*,[43] this 2013 paper was to define the field of aging. Though this paper is supposed to be a summary of aging, it turns out to be a summary of some of the properties of cellular aging. So, to summarize the authors' conclusions, there are three types of hallmarks of aging:

- Primary Hallmarks (causes of damage): genomic instability, telomere attrition, epigenetic alterations, loss of proteostasis
- Antagonistic Hallmarks (responses to damage): deregulated nutrient sensing, mitochondrial dysfunction, cellular senescence
- Integrative hallmarks (culprits of the phenotype): stem cell exhaustion, altered intercellular communication.[40]

This is intended to mean that **genomic instability** — DNA damage at all levels, from simple single-base mutations to aneuploidy (having too many or too few chromosomes) — is a "cause" of aging. **Telomere attrition** is a normal part of the cell cycle, and these protective "aglets" of the chromosome (telomeres) are shortened during each cell division, one telomere shortened on opposite ends of each double-stranded chromosome on each chromosome as they are never fully copied every time the cell divides. An enzyme, telomerase, controls the length of the telomeres, but is only present in embryonic cells and stem cells (though its activity declines with age). This turns out to be not as sure a timer as the original theories, which equated telomere attrition with the Hayflick Limit (the maximum number of divisions an individual cell can undergo before it stops being able to divide — replicative senescence). Before Leonard Hayflick discovered a limit to the number of times a cell can divide, Alexis Carrel, the Nobel Prize winner who first developed cell culture, "proved" that cells were immortal, but it was later found he continuously fed young fibroblasts, an easy cell to grow, into his cultures. Past this limit, either cells won't divide or if they do, one or both "daughters of division" (a great name for an all-girls band) will die.

The **epigenetic alterations** are something we will spend considerable time on, as it is agreed upon by all that epigenetic changes correlate with age changes (and, as believed by some, with the probability of continued life, as we'll discuss) — but the main thesis here is that aging occurs at the cellular level: the accumulation of DNA lesions and the attrition of telomeres, a means of measuring age in cell divisions as with the paramecia we discussed. The epigenetic variations discussed are largely the result of "epigenetic drift" which occurs with age as epigenetic mechanisms become faulty.

Loss of proteostasis refers mainly to the unfolded, or incorrectly folded proteins that accumulate with age. Mostly the lethal effects of these proteins may be appreciated in the aggregation of unfolded proteins that produce amyloid deposits in aging tissues. The cause of this is not known, but later we'll encounter one hypothesis.

The "Antagonistic Hallmarks" are not, as the authors indicate, "a response to damage" (which on the cell's part might include repair), but rather the immediate results of damage. **Cellular senescence** "results" from telomere attrition (and additional forms of senescence result from other causes such as overexpression of oncogenes, i.e., "cancer genes". The presumption that **stem cell exhaustion** (one of the "Integrative Hallmarks") results from telomere attrition is surprising as stem cells should be able to self-renew and provide most other cell types and should have active telomerase that keeps their telomeres long — but they too age and lose that ability. Others become senescent or perhaps even cancerous and die by apoptosis, or some other way of cellular suicide or even simple necrosis — they simply die.

Certainly, aging correlates with epigenetic changes, as age can be measured, as we will discuss, by DNA methylation (DNAm) changes using a DNAm "clock", an AI process that computes the fraction of a select (informative) sample of the tens of thousands of sites in vertebrate DNA that include the dinucleotide CpG in which the "C" (cytosine residue) has the probability of adding an additional methyl group attached to it to make 5-methylcytosine, a signal that may block DNA availability to transcription factors and repair enzymes, but that can also be positive in some cases. Assessing the proportions of particular "informative" methylated CpG sites accurately determines the age of the animal from which the DNA was taken, but we'll talk more about this later.

Mitochondrial dysfunction basically means that the mitochondria become less efficient at converting the proton motive force (provided by NADH) that drives the mitochondrial machine (protons pumped into the intermembrane space trying to push their way back) into ATP, and so they decrease ATP production, but also increase the number of ROS produced per oxygen molecule consumed. So, there are two problems: less energy produced per food molecule taken in, and more ROS production, with potential damage.

The "Integrative Hallmarks" are the final results of the prior two stages, so that telomere attrition and genomic instability might both lead to cellular senescence, and both or either may lead to **stem cell exhaustion**. The last category, **altered intercellular communication**, has many age-related effects — such as Wnt signaling being discovered to be defective in muscle satellite cells[44] — but what López-Otín et al. were primarily talking about was inflammation, the production of inflammatory cytokines (a broad category of small proteins important in cell signaling) and the effects these cytokines have on the body.

However, that tells many stories and gives many causes for many reasons, but is it possible that a single cause underlies all of these apparent causes and effects of aging?

We already know that in *C. elegans* (López-Otín et al. are mostly concerned with vertebrate aging) *there is* a single cause of aging behind all the disparate "hallmarks", and that cause is the loss of "resilience" (also referred to as organ reserve or vitality), the all-cause source of cellular protection; but what is that "resilience" and what happens to it? And what do we know about "resilience" or "organ reserve" or "vitality"? Clearly, "resilience" means the ability to rapidly recover from difficulties, toughness, while organ reserve suggests possessing more than what is needed to overcome difficulties — vitality being similar. So, what we are talking about is that all members of the same population have the same environment and the same genotypes, yet some individuals succumb to whatever damages occur due to their environment — we should assume those individuals lack "resilience".

Now, the other thing we know about resilience is that the more stress an organism faces over time, the more quickly it loses its resilience, such that incidents easily survived in youth are lethal when resilience has sufficiently declined. Conditions that decrease "stress", such as placing antioxidants in *C. elegans* media, seem to slow the loss of resilience. However, perhaps it is those conditions where repair systems are absent, like the lack of heat shock proteins and DAF-16 (a FOXO transcription factor which when activated will enter the nucleus and turn on many repair and maintenance functions), that tells us more.

Thus, the more damage, the faster "resilience" is lost, and the shorter the lifespan is, unless that damage is repaired. So, to shorten lifespan, we can either raise the level of damage or lower the level of repair. We can assume that each organism would receive the same amount of damage (in the case of the HSP-1 and the DAF-16 mutations, both being transcription factors that unleash repair), so it's not the damage that results in (for example) a shortened lifespan, but whether or not that damage is repaired. So what might this "resilience" be? It turns out that another theory of aging is called in to account for that; and as we will see, this more basic understanding of why cells die illuminates not only what resilience is, but also how it relates to the *real* "Life's Timer", which Neill guessed, but later dismissed: the circadian clock.

The redox theory of aging

The subject of redox (reduction and oxidation) is something many forgot from high school chemistry. There we used an acronym for oxidation, LEO — Loss Of Electrons, oxidation — to which I always added GER — Gain Of Electrons, reduction, as the lion (LEO) says "GER". Now we've basically talked about this: all of life's energy source is high energy electrons (electrons that would rather be somewhere else, i.e., unstable) moving from a place of high potential energy to a place of low potential energy — just as in a battery, and just as in a battery that loss of potential energy can be converted into work.

In aerobic animals, most of the high potential energy electrons are transferred to oxygen molecules to form water — a process called oxidative phosphorylation. This process takes place in the mitochondria and is mediated by the hydride ion acceptor NAD^+ which then becomes NADH. NADH is a dinucleotide (meaning it is two nucleotides linked together — DNA and RNA are polynucleotides, i.e., many nucleotides linked together), but with an unusual diphosphate bond connecting them rather than a single phosphate linkage (see Figure 15). NAD^+ acts as a prosthetic group (also called cofactor) for a class of proteins, particularly *dehydrogenases*, receiving hydride ions and ultimately passing them to a chain of four multiprotein complexes (complexes I – IV) with metal-sulfur prosthetic groups to accept these high energy electrons and pass them to the next complex in the series. Finally, a fifth group, though embedded in the inner mitochondrial membrane, just into the mitochondrial matrix, uses the chemiosmotic potential generated by the four complexes of what is called the *electron transport chain* (ETC) to manufacture ATP from inorganic phosphate and ADP.

We know that several enzymes use NAD^+, the energy-poor form, for other purposes than electron transport, and we'll learn that this can be a problem when supplies are limited: the main function of NAD^+ is to pick up hydride ions and pass their high energy electrons and attendant protons (the normal hydrogen atom consists of one electron orbiting one proton, and the hydride ion has one proton and two electrons) to the mitochondrion to produce the energy the cell needs for all of its functions. The mitochondrion does this with a series of metal-coordinated proteins (proteins with metal ions in their center, such as iron and copper) that form a series of pathways for high energy electrons to travel across the inner mitochondrial membrane, dragging a proton with them, and such protons can be enticed into the mitochondrion's inter-membranal space against the pressure of their own positive charges, becoming crowded together.

Once all these protons (hydrogen ions) are all crowded together, they exert a positive pressure. Since all protons repel each other, the closer they are, the stronger the repulsion, so filling up the inner mitochondrial membrane with protons is like blowing up a balloon; if you are holding the neck of the balloon clamped, and then release it, it develops a force that may propel it across the room. You could even imagine preventing it from moving, so when unclamped, while holding the balloon in place, it will produce a "wind" that can turn the blades of a pinwheel or a turbine to produce electricity, right? The potential energy of the filled balloon depends on the pressure within the balloon. When you're blowing it up, it takes energy, and you are fighting against the force of the air pushing outwards. Protons store even more energy, because they are all positively charged so they exert a much stronger opposing force when trying to shove them together.

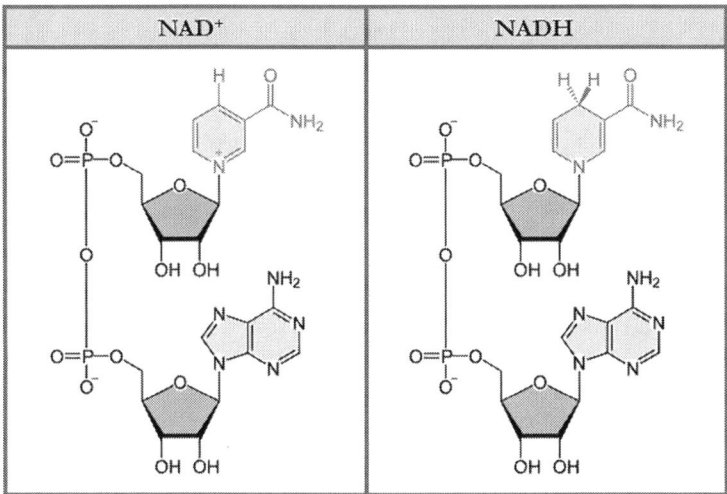

Figure 15: In this diagram, adenosine is the bottom nucleoside made of a sugar (called ribose), colored dark gray, and a base, adenine, colored light gray. The top part is the other nucleoside, also made of ribose (dark gray) and another base, called nicotinamide (light gray with gray lines). The top and bottom nucleosides are connected by a diphosphate bond, forming the dinucleotide. Adapted from "Energy in living systems: Figure 1" by OpenStax College, Biology (CC BY 3.0), https://creativecommons.org/licenses/by/3.0/us/.

As energy is equal to the force applied over the distance, pushing protons into the intermembrane space requires a force applied over a distance — which means that energy is being stored as a gradient of hydrogen ions, as a higher concentration in the intermembrane space than in the inner mitochondrial matrix is formed. The electrical energy stored in the mitochondrion, in terms of protons concentrated in the intermembrane space, use the mutual repulsion (the "proton-motive force") to drive a molecular turbine called ATP synthase (it actually turns) that adds an inorganic phosphate group (PO_4^{3-}) to ADP (adenosine **di**phosphate [two]), to make ATP (adenosine **tri**phosphate [three]), a high-energy acid anhydride.

ATP is the gasoline that drives all cellular "engines" — protein molecules that *do* something (enzymes), such as muscle protein complexes (actin-myosin) that contract allowing us to move, or some that act as molecular pumps that drive ADP, ATP and a host of small molecules across cellular membranes. So, ATP drives our muscles and molecules. One curious thing is that you can see if the mitochondria of a cell are active or not with electron microscopy: the active ones are puffed out, and the inactive ones sort of look shrunken.

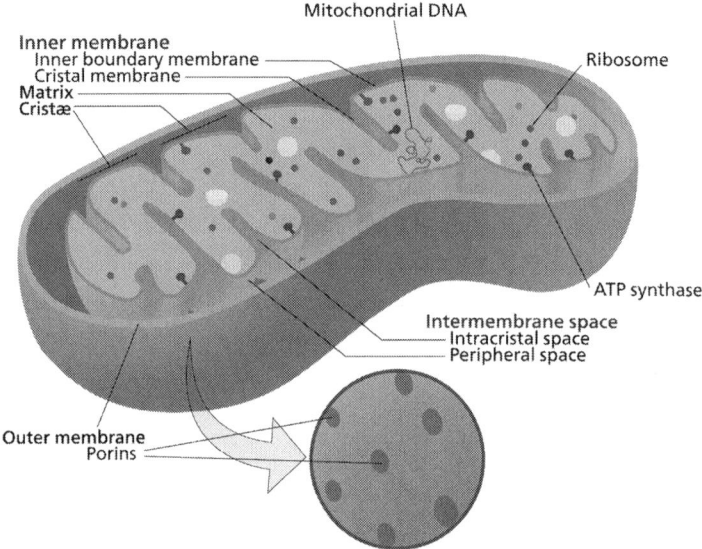

Figure 16: Mitochondrion structure. Adapted from Kelvinsong; modified by Sowlos, CC BY-SA 3.0 <https://creativecommons.org/licenses/by-sa/3.0>, via Wikimedia Commons.

So, Figure 16 shows an illustration of the basic structure of the mitochondrion with two impermeable "bags" of membranes, one of them a smaller very wrinkled bag (forming cristae — finger-like projections of the membrane to increase surface area), and the other one a smooth outer bag. The space separating these bags is the intermembrane space, and it is that space that gets filled with protons as the high potential energy electrons pass down the complexes of the electron transport chain (ETC) — I, II, III and IV — embedded in the inner mitochondrial membrane, losing potential energy at each step. The lost potential energy is used to pump protons into the intermembrane space against the concentration gradient. For each electron transported along the ETC, about 10 protons are transported to the intermembrane space. It is along this pathway that electrons are passed as pairs to oxygen molecules to form water — but also, when single electrons are passed to oxygen molecules to form the superoxide anion radicals.

Eventually, the proton-motive force generated by the higher concentration of hydrogen ions in the intermembrane space becomes great enough to create a flow of protons to turn the turbine that is the F_0F_1 complex (complex V), called ATP synthase, which uses mechanical-chemical energy to pop together an ion of inorganic phosphate (a normal constituent of soda pop) and a molecule of ADP to form the high energy acid anhydride bond, that connects the three phosphate groups and so stores considerable potential energy as ATP. The ATP/ADP cycle is shown in Figure 17.

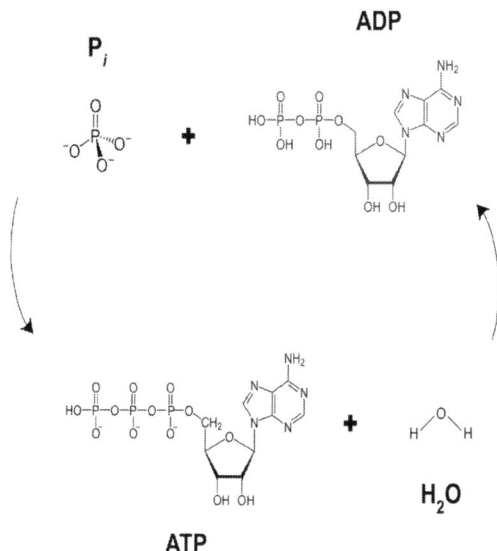

Figure 17: Conversion of ADP and phosphate into ATP and vice versa. Adapted from Butrboy, Public domain, via Wikimedia Commons.

So, the idea is that relatively simple energy-rich nutrients are delivered to the mitochondria which essentially burns them into carbon dioxide and water — gaining a huge increase in energy as compared to anaerobic processes, as each molecule of glucose (a simple six-carbon sugar) burned by the anaerobic cell gives rise to two molecules of ATP, while that same molecule of glucose, when metabolized aerobically in the mitochondria yields 30 molecules of ATP. It's a fifteen-fold gain in energy, so it is clear why cells that incorporated what was believed to be endosymbiotic alphaproteobacteria (that gave rise to today's mitochondria) succeeded so well.

However, as has been noted by ecologists (and the basic principles of thermodynamics), "There's no such thing as a free lunch." As I said before, between 0.1% and 3% of the oxygen atoms that are processed by the mitochondrion become superoxide radical anions and have significant effects on lifespan — and as usual in biology, a complicated one. Do note however that the ETC becomes less efficient with aging, producing fewer ATP molecules per glucose and producing more ROS as well.

So, let's simplify it all a bit — here is a sort of very basic description of what happens.

A cell takes nutrients (and for the sake of simplicity we will consider only one at present, glucose, the six-carbon sugar that our cells run on). First, that glucose is broken apart and oxidized, and at the end of a process called glycolysis, we are left with a highly oxidized three-carbon molecule called pyruvate (pyruvic acid) that is further degraded and combined with a molecular "handle" (coenzyme A) to become acetyl-CoA, which then joins a merry circle dance of molecules exchanging partners (the tricarboxylic acids cycle, or TCA cycle), with carbon dioxide being formed, as well as ATP (actually GTP, but they're energetically equivalent and can be interconverted, as GTP + ADP ATP + GDP), the high energy electron carriers NADH (which carries a pair of electrons) and $FADH_2$, which holds a single high energy electron. These are the electrons that are transferred to the mitochondrial ETC to be converted into ATP energy.

Like electricity in the modern home, ATP, in the cell, powers most processes. So, the eukaryotic cell has the ability to use high-energy molecules in its environment to produce huge amounts of ATP, much more than needed for simple life maintenance.

In many ways, the eukaryotic cell's discovery of the mitochondrion and oxidative phosphorylation was analogous to humanity's discovery of fire — but as useful and powerful as fire is, it has a sooty side. It does damage as well, and that inevitable damage must be dealt with. In the case of cells, that damage results from the generation of ROS (reactive oxygen species), which cause oxidation damage, and the generation of high energy nitrogen compounds as well.

We saw that the apparently unintended production of the superoxide radical anion resulted from faulty reactions in which a single electron changes the oxygen molecule to the highly reactive superoxide radical anion, a species which can do considerable damage, especially as it's stuck within the mitochondrion, which will not allow charged particles (all anions are negatively charged as compared to cations, which are positive) to cross its inner membrane. To address this problem, the mitochondrial enzyme superoxide dismutase converts the superoxide radical anion into the less reactive hydrogen peroxide, a molecule that readily diffuses through the mitochondrial membranes to enter the cytoplasm, causing hydrogen peroxide (and nitrogen oxides) to damage important biomolecules.

The damages caused by peroxides and other ROS must be repaired lest the cell degrades. It is agreed by most biologists that the ultimate cause of cellular death in cellular aging is oxidative stress (an overabundance of ROS) — and eukaryotic cells have several systems dedicated to repairing the redox damage done by ROS (as well as many other sorts of damage). In fact, eukaryotic cells in general, and ours certainly, have two interdependent systems to repair oxidative damage, and many enzymes unfamiliar to many are involved in the repair of such oxidative stress. Both systems are based on a simple molecule, a variant of the dinucleotide NADH "marked" (as this mark doesn't affect the redox potential of NADH, but is there for different reasons) with an attached phosphate, and called nicotine adenine dinucleotide phosphate, or NADPH.

NADPH is like NADH, a hydride ion donor, but it has another function as well (one of even more importance to life than even fixing oxidation damage), which is growth and maintenance. NADPH provides the hydrogen anions for the reductive synthesis of molecules vital to the continuance of life; it is used as a source of reducing power. One part of the energy life extracts from the high energy electrons of NADH and $FADH_2$ goes into maintaining life's quotidian activities — providing energy for movement, for pumping ions, producing stomach acid, etc. In this case, those two electrons carried by NADH will be passed, through the electron transport chain (ETC), to water with a loss in potential energy enough for the creation of three ATPs for each NADH that transfers its electrons to the mitochondrial ETC. However, there is more to life than simply moving; growth, maintenance and repair, and it is in pursuit of these goals that NADPH uses its high energy electrons.

Refresher: catabolism and anabolism

The ancient Taoist symbol of yin-yang is commonly known to look as shown in Figure 18.

Figure 18: Representation of the yin-yang symbol. Adapted from Iruka13, Public domain, via Wikimedia Commons.

In Taoism, which I haven't mentioned — a Chinese religion and philosophy based on eternal cycles — immortality is assured; every death is the beginning of a new life. The symbol itself represents the two opposites that work together to create the whole. So, there is no darkness without there being light. No dry without wet, etc. The white section represents yin, and the black, yang. As it was originally conceived as two fish constantly circling each other, let's take the analogy to metabolism.

So far, we have discussed one half of metabolism, the part called *catabolism*, the breakdown of complex molecules to yield energy. Let's think of that as "yin": here, complex molecules, containing chemical potential energy, are broken down to provide NAD^+ with the hydride ions to become $NADH$ (that is, to pass their chemical potential energy to create $NADH$), and that energy is liberated as electrons flow down the ETC. In each subsequent "link" on the "chain" of the four complexes, the electrons have an increasingly lower potential energy, as they "lure" their accompanying protons into the mitochondrial intermembranous space to produce the ATP an organism needs to escape enemies, or search for prey or a mate. In this process, the cell liberates energy from the environment by breaking apart complex, energy-rich molecules ("food"), and capturing their energy and atoms and "specialty molecules" (like vitamins) for their own needs. However, that's only staying alive — but there is more: reproduction, growth and maintenance.

While during catabolism NAD^+ accepts hydride ions, acting as "mules" to carry them to the ETC where they will ultimately produce ATP to drive cellular processes, in anabolism $NADP^+$ is the hydride ion acceptor, forming NADPH, and while this molecule has exactly the same reducing potential as NADH, it has totally different functions. While much of the NAD^+ is locked away in the mitochondrion, a cellular compartment (neither $NAD(P)^+$ nor $NAD(P)H$ can pass through cell membranes), NADPH functions in the cytoplasm (only 10% to 15% is found in the mitochondrial fraction) as a source of reducing potential for synthesis and repair of the oxidation damage incurred during catabolism. So, the yang to catabolism's yin is *anabolism* — which is, as in the Taoist symbol, exactly the reverse of catabolism: in anabolism, the energy obtained by the cell (from catabolism) is used to build complex, energy-rich molecules.

So, at this superficial level, the yin-yang analogy is correct; you cannot build complex molecules (anabolism) without the energy and materials provided by catabolism, and similarly catabolism requires the complex molecules, like enzymes, produced by anabolism. So, one "fish" drives the other. While the energy captured by NADH is used to produce ATP to provide for the quotidian and immediate needs of the organism, the reducing power of NADPH is used to add electrons in synthetic reactions. For example, in order to synthesize DNA, the RNA nucleosides based on the sugar ribose must have their ribose moieties reduced to deoxyribose, because that is what is required to make DNA, and NADPH is the enzymatic cofactor that provides the reducing power for this reduction.

Where does $NADP^+$ and NADPH's reducing power come from?

NAD^+ is the immediate precursor of $NADP^+$; the enzyme NAD^+ kinase (a "kinase" is an enzyme that adds a phosphate group to another molecule) works on NAD^+ and not so well on NADH (the "reduced" form). The difference is important because there is a decrease in NAD^+ with age and a decrease in the $NAD^+/NADH$ ratio over time. This results in part from the destruction of NAD^+ by several enzymes that use it as a substrate. This causes a decrease in the $NADPH/NADP^+$ ratio as well, as NADPH is oxidized in order to repair damage due to oxidative stress. The decrease of $NADPH/NADP^+$ and particularly of the ratio of oxidized to reduced glutathione (another antioxidant biochemical molecule) results in a decrease in the reduction potential in the cytosol and the nucleus. The change in redox potential may profoundly affect enzymes with reactive thiols (-SH groups, cysteine and methionine are the only amino acids with this group).

While the majority of the hydride ions that attach to NAD^+ come from dehydrogenase enzymes used in glycolysis (enzymes that use NAD^+ as a cofactor) and in oxidative phosphorylation in the mitochondria, the reducing equivalents gained by $NADP^+$ come from several different pathways. The *pentose phosphate pathway*, or PPP, is the best known pathway for the reduction of $NADP^+$, and it is an alternative pathway for the oxidation of sugars that are not headed for mitochondrial energy conversion but for anabolic synthetic reactions. The sugar ribose is one important product of this pathway. In these reactions, two dehydrogenase enzymes that remove the hydride ions use $NADP^+$ as the hydride acceptor and produce NADPH as a product, and three other cytosolic enzymes and five mitochondrial enzymes also use $NADP^+$ cofactors as hydride acceptors.

Aging, death and energy

As stated, the other vital purpose served by NADPH is the repair of the oxidation damages incurred in energy production. It's actually not until I learned the "intricacies" of the redox stress theories of aging that I understood that it must be involved in cellular aging, as there is a constant loss of cytosolic (the non-compartmented parts of the cell's cytoplasm, not including cytoplasmic membrane-bound organelles) reducing potential with age, as well as a constant gain in reducing potential in the endoplasmic reticulum which is normally kept at a high oxidation potential for protein folding, which requires oxidizing compounds to form the intermolecular and intramolecular disulfide bond (-S-S-) that often determines the final shapes (and functions) of proteins and holds together protein complexes.

We see that this very much looks like entropy at work, with cellular compartments like the cytosol and the nucleus starting in youth with highly reducing environments, with the majority of the NADPH and glutathione reduced. Glutathione (GSH) is a tripeptide (three amino acids joined), one being the amino acid cysteine with a thiol group (-SH) that can easily be oxidized. This tripeptide is present at very high, millimolar concentrations, while NADH and individual proteins are at micro and nanomolar (and less) concentrations.

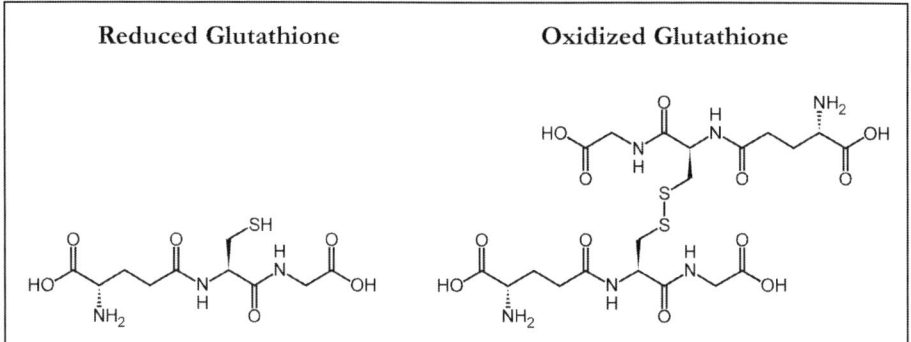

Figure 19: Molecular structure of the reduced and oxidized forms of glutathione (GSH).

With its high concentration of thiol groups, reduced glutathione offers a vast and ready source of reducing equivalents. Note that it takes two electrons to reduce oxidized glutathione. The ultimate source of GSH's reducing power is NADPH and the enzyme glutathione reductase, which transfers the hydride ion to reduce glutathione. There are a number of enzymes that utilize reduced glutathione as an electron donor, such as the glutaredoxins, important enzymes that reduce molecules oxidized by ROS produced in oxidative phosphorylation. However, there is another parallel repair system in which NADPH-provided hydride ions are passed to a family of redox repair enzymes called thioredoxins (also with an important thioredoxin reductase that depends on NADPH). The members of this family that reduce peroxides are called peroxiredoxins (fun to say), and are among the most common enzymes in the cells.

Aubrey de Grey once wrote to me (amidst a months-long argument) that Nature selecting both lifespan limits and fitness would be like stepping on the brake and the accelerator at the same time. This is simply not true, as I'll explain (as different life-stages have different phenotypes), but it does sound true at first.

So, if we have a set of oxidation reactions, destroying biomolecules to liberate energy, as well as reduction reactions using the liberated energy to create biomolecules, isn't that like stepping on the brakes and accelerator at the same time? There are, of course, solutions to that problem; one is to physically separate the oxidation reactions for energy retrieval from the reduction reactions for macromolecular synthesis and the reduction of oxidized cellular molecules. This is actually done, as the mitochondria produce

ATP by catabolizing foodstuffs, and the rough endoplasmic reticulum, cytosol and the nucleus synthesize new molecules based on the energy (ATP) supplied by the mitochondria to manufacture nucleoside triphosphates, whether ATP, CTP, GTP, UTP or TTP — each with the same chemical potential energy.

During DNA and RNA synthesis, for each nucleotide being added to a growing DNA or RNA chain, one ATP energy equivalent is lost. During protein synthesis, for each amino acid added to a growing protein chain, about three ATP equivalents are required. Every cell movement, including the movement of its flagella or cytoskeleton, requires the breakdown of ATP (ATP → ADP + phosphate + energy). The energy expended in growth and maintenance must be provided by catabolism in the mitochondrion (and to a lesser degree in the cytosol).

The synthesis that is required for growth and repair is limited to the energy provided by catabolism (conservation of energy holds). The fact that energy decreases with age is clear to the elderly — in fact David Neill proposed that the lack of energy was responsible for aging. But why should that be? It is almost as though a cell or the organism was like a battery, charged with a defined amount of energy that eventually runs out. However, the cell's or body's energy is determined by its intake of food, so how is that possible? If more energy is required, then why not simply eat more?

What is noticeable about the aging cell — the cells of older organisms that lack Stroustrup and Fontana's "resilience" — is that they also lack NAD^+ (while in young organisms NAD^+ predominates, in the cells of old organisms the reduced form NADH predominates). And while in young cells NADPH is predominately in its reduced form, its oxidized form increases in the cells of old animals.[45] A steady age-dependent decrease in the organ contents (meaning within the cells of that organ) and in plasma levels of GSH is a recognized marker of aging[46] and is associated with multiple morbidities. It is our contention that the loss of energy generating power by mitochondria, together with its increased production of ROS which increasingly oxidize the cytosol and nucleus, and the increasingly reducing environment of the endoplasmic reticulum, are in sum the "resilience" that is lost with aging and the cause of all of the "hallmarks" of aging. But how and why do these changes occur?

The circadian rhythm and sleep

In Latin, *circa* means "about, nearly, close to", and *dia* means "day", so *circadian* means "about a day". It is a known fact that all higher vertebrates have a built-in rhythm such that the times of activity and the times of rest and sleep are predetermined for the species. It is also known that disrupting this schedule shortens lifespan.

Why do some mammals spend half their day (like dogs), sleeping? This is not a peripheral question, because sleep is universal in vertebrates (and analogous states exist in invertebrates as well). In order for an organism to spend a significant fraction of its life in a state that does not contribute to its intake of energy and materials or help it reproduce, and further leaves the animal vulnerable and immobile, this state must be of extreme importance. We will describe sleep as a global phenomenon, but this function is so basic to neurons that even simple neural networks grown in vitro show sleep-wake cycles. Recent work by Steve Horvath convincingly shows that it's in the brain, and in no other organ tested, where the genes that establish the circadian rhythm are hypermethylated with aging (meaning that the epigenetic aging markers related to the circadian rhythm are found in the brain).[47] The functional connections between changes in sleep patterns and a general dissipation of circadian rhythms, and age, are clear.

We discussed that the separation of metabolic processes that might interfere with each other — physical separation by impermeable barriers, i.e., "compartmentalization" — was part of the solution, but these processes are also separated temporally, and the astute reader may have already recognized that sleep — a time of rest (night in humans, day in rats) — would be a time when extensive ATP synthesis might not be needed for ongoing activities, and so it would be a good time for growth and reparation. Shakespeare himself recognized this when he wrote "sleep that knits up the raveled sleave of care".

The circadian clock is present in every human cell, and yet it's systemically controlled by a region in the brain's hypothalamus called the superchiasmatic nucleus (SCN). At the organismic level, the hypothalamus periodically releases hormones (actually hormones that release other hormones in the pituitary gland — the so-called "releasing hormones" in the hypothalamus) which control timing in other organs. In a recent study, Cai Dongsheng of the Albert Einstein College of Medicine showed that in the aging hypothalamus, lowered levels of the releasing hormone GnRH (gonadotrophic releasing hormone) failed to entrain other tissue clocks resulting in aging; supplementation of GnRH significantly extended the lifespan of mice.[48] It is this clock, present in the cells of the SCN, that controls its activity.

At the cellular level, what is clear is that there is a complex system called a transcription-translation feedback loop (TTFL) that determines when the catabolic reactions resulting in ROS production and consequent oxidative stress, and then when the synthetic and reparative use of reducing equivalents, take place. Simply stated (the loop has other subloops we won't be bothered with as they are peripheral to the story), two proteins — Bmal1 and Clock — form a dimer that acts as a transcription factor to cause the transcription of the cryptochrome (CRY) and period (PER) proteins, which are translated in the cytoplasm (as is normal), eventually forming PER-CRY heterodimers that travel back to the nucleus where they prevent their own transcription by inhibiting the transcription of the Bmal1 and Clock proteins.

Eventually, PER-CRY is destroyed and the cycle begins again. However, this clever mechanism seems imposed on deeper rhythms of oxidation and reduction found to exist even when the TTFL mechanism is disabled — it was found that the oxidized state of peroxiredoxins (a family of antioxidant enzymes) oscillates with a circadian rhythm in human red blood cells, which having no nucleus cannot contain a TTFL. According to Sandipan Ray and Akhilesh B. Reddy, researchers from the University of Cambridge, "Subsequently, oscillations of peroxiredoxin proteins (PRX) have been established as evolutionarily conserved markers of the clockwork, pointing to redox cycles as a likely unifying principle among disparate organisms."[49]

The redox state of the cytosol and the nucleus greatly affects the many enzymes that have reactive thiol (-SH) groups present on cysteine and methionine residues (amino acids in a protein-chain), changing many properties of the enzymes, including activity, protein-protein binding and protein-nucleic acid binding.[50] Among other enzymes, a particularly interesting one is telomerase — the enzyme that effects the elongation of telomeres following cell division. Telomerase functions under reducing conditions, but not in an oxidizing environment.

It is easy to imagine that other enzymes will change their properties (in a way similar to aging changes) with an increasingly oxidizing environment. In fact, the binding of the Clock-Bmal1 dimer to DNA depends on the ratio of $NAD^+/NADH$ (the cell's "redox potential"). While the redox status controls the "clock", conversely, the "clock" controls the redox status, as many of the antioxidant enzymes and cofactors and ROS-responsive genes are controlled by the "clock". Please note that as the ratio of $NAD^+/NADH$ controls Clock-Bmal1 binding, there is a direct connection between the redox state and the circadian clock.

In the fruit fly, *Drosophila melanogaster*, about two dozen neurons control sleep. These neurons are controlled by the activity of an enzyme (actually a potassium channel — an opening in the center of this cell-membrane-spanning-protein that allows a potassium ion current to flow into or out of the cell) with an NADPH cofactor that is exposed to the oxidative by-products of mitochondrial metabolism; once ROS have oxidized that cofactor to $NADP^+$, the channel shuts, thereby raising the action potential of these neurons and inducing sleep.[51]

Reciprocally, clock genes control many metabolic processes related to energy production and gene control — the Clock protein along with Sirt1 (sirtuins are a family of protein deacetylases, meaning they remove the acetyl groups [$-CH_2COOH$] from proteins). This is of particular interest to us, because when those acetyl groups are added to chromatin — to the "histone proteins" that form the nucleosome hockey pucks around which our DNA is wrapped — that "opens up" the structure of chromatin, allowing its genes to be transcribed into mRNA (messenger RNA). However, in old age, tracts of DNA that should be de-acylated are not, resulting in aberrant gene expression (genetic misregulation) in the cells of older animals. The loss of NAD^+ is responsible for that, as NAD^+ is required for deacetylation.

As stated, there is a loss of NAD^+, GSH and NADPH in aging cells, but why? We know the circadian clock controls many aspects of the metabolism of these important stores of reducing equivalents; it controls the NAD^+ salvage pathway, where NAD^+ is used as a substrate by sirtuins and cleaved in the process, and the salvage pathway restores NAD^+ to its original form. If that weren't enough, even the de novo synthesis of NAD^+ requires the enzyme NAMPT (nicotinamide phosphoribosyltransferase) which also varies rhythmically. The enzyme NAD^+ kinase, which converts NAD^+ to $NADP^+$, is rhythmically controlled. The central metabolic enzyme that produces the acetyl-CoA that powers the mitochondrion is also rhythmically controlled. Also, SIRT3 in the mitochondrion is controlled by the clock, and controls the rhythms of mitochondrial functioning. As we know clock functioning is disrupted by aging (and disrupted clock functioning also results in shortened lifespans), loss of a robust circadian rhythm during aging may account for part of the energy-loss problem.

Sleep has myriad functions; for example, it removes potentially toxic substances from the brain (which actually shrinks as these by-products of its metabolism are secreted via cerebrospinal fluid into the glymphatic network — which acts as the lymph vessels of the brain). It is during sleep that memories become consolidated. During the early phases of sleep, NADPH is used to reduce ribonucleotides to deoxyribonucleotides for DNA production, for the production of fats (cholesterol) and for many other synthetic purposes (anywhere a synthetic reduction must take place).

However, during the latter part of the sleep cycle, the one major purpose appears to be the restoration of cellular redox potential and the repair of oxidative damage, basically bringing the body's cells back to the state they were in the day before, being "refreshed" as it were, but here's the thing: that doesn't happen. Every day, the cytosol and nuclear compartments grow more oxidizing, while the endoplasmic compartment grows reducing (which probably results in improperly folded proteins resulting in the unfolded protein response). Why doesn't the cell allocate the energy necessary to fully recharge itself? The daily loss of cytoplasmic and nuclear reduction potential, and the increase in the oxidizing potential of the endoplasmic reticulum accumulate and become a measure of aging. The reducing potential (let's just stick to that) of the cytoplasm declines as the ratio of NADPH/$NADP^+$ and the amount of GSH decreases. It is known that disruptions of the circadian cycle affect both aging and neural degeneration, but why?

This was always a problem. Why does the cell become oxidizing? What happens to its reductive potential? Why does GSH go down with age? Why isn't $NADP^+$ reduced by the PPP (the metabolic pathway that consumes glucose and generates NADPH) and its various other routes?

As we'll discuss, many of these things are tied together by the age-related down-regulation of some, but not all genes (and the age-related up-regulation of others, especially concerned with inflammation).

While protein synthesis in aging cells slows down, and while there is a global decrease in protein production, the majority of proteins are produced as they normally would be. However, some proteins are down-regulated — some of them are very special, such as gamma glutamyl-cysteine synthase (the enzyme that produces GSH), various DNA repair enzymes, important metabolic enzymes such as G6DH (the first oxidative enzyme of the PPP that charges NADPH), and enzymes for the repair of oxidation damage including the thioredoxin and thioredoxin reductase enzymes.

Despite the cell's requirements for additional energy, cofactors and antioxidants, the cell loses the ability to produce them. And the consequences of this are that NAD^+ is used up by sirtuins in deacetylation reactions, polymerized by PARP (poly ADP ribose polymerase) to form the poly ADP chains used to mark DNA damages, and consumed by CD38. CD38 is a protein found on lymphocytes, which can act either as an activating receptor, causing T-cells (T-lymphocytes) to be activated to produce a variety of cytokines, or as an enzyme, producing the signaling molecules ADP-ribose and a little of cyclic ADP-ribose. It takes 100 NAD^+ molecules to produce one ADP-ribose molecule.[52] The conversion of NAD^+ to $NADP^+$ by NAD^+ kinase is also slowed down as levels of the NAD^+ fall.

The complexity of all of these reactions and their side effects are still not completely known. For example, PARP1 — the enzyme that forms the poly-ADP chains that mark DNA damage — also regulates clock functions, while the sirtuins (SIRT3) control the rhythms of mitochondrial respiration. We have not even discussed the cell's response to these conditions. Normally we would discuss the transcription factors (FOXO, HSP1) that unleash a host of reparative enzymes to fight the problems of oxidation and other damages, but if we know as much as we do — if we assume that increasing the levels of GHS, the $NAD^+/NADH$ and $NADPH/NADP^+$ ratios, and getting rid of ROS should extend life and this seems to be the case — we can take a different approach.

The work of David Sinclair[53] and Leonard Guarente[54] established (in mice) that somehow tricking the animal's cells into producing more NAD^+, or substantially reducing ROS production, extends lifespan to an extent. Living another 30% longer would certainly be a boon — especially as that would delay and ameliorate the diseases of aging, which are a tremendous cost to an aging population, but here's the thing: as we saw in the experiments of Stroustrup and Fontana, such solutions proportionally extends all life stages. So, if such treatments are successful, what we will extend is old age.

This resembles the legend of the goddess Eos, who loved the youth Tithonus, and whose pleas to her fellow gods to make him immortal were granted — but what Eos neglected to ask was to give Tithonus immortal youth as well, and he grew older and uglier with time until Eos just locked him away from sight. That is not a good plan, as what we're after is immortal youth, though it seems that the classical aging theories will not ever give us this possibility. Sure, you can keep the old jalopy running, but can we do better than that? If I didn't know so, I wouldn't be writing.

Before we see how this can be accomplished, I want to completely change the direction of David Neill's conclusion about "Life's timer" (which I take to be the circadian clock). Dr. Neill writes about his theory of aging that "Rather than a genetic pathway/program, this theory proposes a longevity timekeeper that can delay the rate of accumulative cellular damage during the post-maturational period of life."[26]

I will demonstrate to you that the opposite is true, and I would say about my own theory that rather than a genetic program, an epigenetic program works with the circadian clock-controlled metabolic cycles to cause the accelerating accumulation of cellular damage during the post-maturational period of life, resulting in death at or before the species maximum lifespan.

7

An autobiographical note or two

When I was quite young (about eight years old), I remember lying in bed at night thinking about the "purpose" of life. It all seemed to come down to reproducing and dying. It was then that I first started to really think about aging and death. Subsequently, when I was about 14, my mother came down with colon cancer — and for four horrible years I watched my mother wither away in such pain that she prayed for death, which she finally received. How is this life worth living if for all we do we end like that? This was Gilgamesh's problem when his friend Enkidu died, in the *Epic of Gilgamesh*, discussed in the beginning of this book.

I finished high school, and after way more time than anyone should take, I finished college — I was not a good student and it was the 1960s, and there were plenty of distractions of the pleasant kind. Yet, after a spell as a junior high school teacher in Kingston, New York, I realized that as pleasant as teaching was, I felt unfulfilled and decided to go back to get an advanced degree. While my original degree was in physics, I took some courses and

GREs to get a graduate degree in biology — my old reveries on life and death still dominated my thinking. At this time, I was married with one child, Danny (who was the Father of Dragons on the *Game of Thrones* series), and another (Rachel) on the way, so I decided on The City University of New York (Herbert H. Lehman campus in the Bronx) as it was convenient and cheap. There, after a while, I worked at the lab of Dr. Susan Wallace, who specialized in DNA repair.

In those days it was "pretty clear" that DNA damage was "the" cause of aging, so my interest in DNA repair was clear to me — and I did some interesting work, finally published in *Biochemistry*. A bit before getting my doctorate, I went to the University of Southern California to work with the fathers of bioinformatics, Michael Waterman and Temple Smith. There I learned to use computers to find out interesting facts about DNA. A very disturbing event, occurred during that time, changed my life. I had a rotten tooth, a back molar that was becoming painful. The dentist, who was also the manager of the on-campus building I lived in, kept putting off examining me. I tried whatever came to my mind to quell this infection, which by now had caused the right half of my face and neck to become swollen. I went back to the "dentist", who said that with that sort of infection, he could not help me — I should see a doctor. I was getting increasingly sick and I called the Kaiser Permanente hospital, told them of my condition and then, feeling really, really ill, took a city bus to the hospital.

Upon arrival, I went to the front desk and asked where to go. I was told that they had no idea, and I should just wait in the ER like others waiting for treatment. At this time, I was getting sicker and sicker — my neck had a bulge like a football, and there were dozens, if not hundreds, in the ER of this big-city hospital. I was about to pass out when I heard my name called. A Dr. Golan, who I can never thank enough, knew I was supposed to arrive at the hospital, and when I did not, he went to find me. I'm not sure how many physicians working at a big hospital would have cared. While he was taking me to surgery, he was painfully injecting what I suppose were painkillers into my neck, but he didn't wait for them to take effect before cutting into my neck. I told him I was going to pass out, but Dr. Golan said no, and the next thing I knew, I was lying in a bed in critical care with a tube going into my trachea and surrounded by lots of machinery with beeps and blinking red lights, and an IV dripping powerful antibiotics into my veins. I was very sick and did not know if I would live until the next day. And that night I had a dream which would change my life.

Nowadays, I rarely remember my dreams, but this was different. I was in an alcove high above a dusky road, lit by a dim red sun. But it was noon — and marching down that road were soldiers, and a person of some importance was addressing the troops, and I knew then that they were marching in honor of me. I knew I was not on the Earth, but perhaps on a planet orbiting a red dwarf star. I knew, perhaps by what the speaker said, that it was 400 years in the future; I can't remember every word of the speech this official gave, but the line that resonated with me and changed my life was: "To honor Harold Katcher, who brought immortality to mankind". And yet it did not feel like a dream, it felt like a vision, a message that I was not about to die, and impossibly more.

It was a literal fever dream; yet, at that point I knew I wasn't going to die. However, it was crazy that I would bring "immortality to mankind". I was already at the point of giving up with the concept that aging could be any more than ameliorated, delayed perhaps, and I even proposed schemes for constructing bacterial DNA repair enzymes containing plasmids (perhaps in liposomes), to help cells to repair themselves, and even wrote up a grant proposal, but nothing came of it.

Nevertheless, by this point I was finished with science and the arrogant scientists I worked with, and I realized that with the then current theories about aging (still current now — the SENS Research Foundation still accepts the idea that damage can be halted or "engineered away" — even though there is an increasingly greater understanding of the importance of redox biology that has changed some opinions, as discussed, but at that time it was still all DNA damage), this research field was trying to do something like sweeping back the tides with a broom. To me, the prospect of getting repair enzymes into all the cells of a body seemed impossible. How to get repair enzymes into all of the trillions of cells of the body? In spite of my dream, I knew there was no way to do more than moderately extend life — diminishing the rate of damage and over-expressing repair factors. Nearly all experts agreed on the proposition that the effort to extend life would be huge and costly, and the degree of possible life extension would not even be worth the cost.

So, as though I wished to make my dream nothing but an impossible dream, I left science to become a computer programmer (though self-taught, I was pretty good). The money in computer programming was much better at that time (the early 1990s) and I loved computer programming (it's still a hobby of mine). I worked for a commercial outfit, doing some routine programming and cleaning up the (horrible) work of a programmer who preceded me — and I'll admit I was happy. I lived in Salt Lake City with my wife Fatemeh and my daughter Sasha. (At the time, my other two children, Rachel and "Dragon" Dan, were already grown and no longer lived with me.) While all too often during the weekends my "beeper" (remember, pre-cell phones?) would go off and I had to spend a weekend trying to debug payroll software, or some other vital function, generally it was a pretty pleasant life.

However, when I received an offer from the University of Maryland to join their overseas program, my wife, who loves traveling, was all for it. I spent a couple of months before my wife and child in Taegu, Korea, at Camp Henry, and when everything seemed comfortable enough I sent for my family. My daughter was a bit past two years old at the time. Taegu, though you've probably never heard of it, is a city of millions in central South Korea — which quite honestly had little interest to me (though I loved Korean food); there wasn't much to do. However, we made many friends, both Korean and American military, so what more do you need than friends? Well, in our rented apartment (very difficult to get in Korea), we could hear the rats running across our ceiling, but they never entered the apartment and we felt safe enough.

After about two years of that — I was a successful and well-liked professor, and I guess as a reward (it was known to the faculty that a Korea assignment was equivalent to "paying your dues") — they sent us (in a C5 transport, with web seating and an empty paint can for a bathroom, sitting next to a tank) to Misawa ("Three swamps"). It was a large American military base in a small coastal fishing town that was much nicer than its name would imply, and to make a long story short (though it could have its own book), I was promoted to full professor and then appointed the Academic Director for Natural Sciences for the Asian division — Japan, Korea, Okinawa (though officially part of Japan, Okinawans don't consider themselves Japanese) and Guam.

Though things were going very well, after having lived in Japan for nearly a decade, my wife decided that she didn't want to die in Japan — and we decided to return to the USA. As I had responsibilities, my wife and child went first, and I soon followed. I returned home to our house in Salt Lake City, and since I was nearing retirement age, I still worked for the University as an adjunct teaching online courses. At that time, I believe, we — UMUC, now UMGC (Global Campus) — had the world's biggest online program.

So, I suppose that would be most people's dream; being able to work from home and still make a decent salary and have interesting work (I taught both physical and biological sciences, though I missed teaching math, which I did in Asia). However, what about that dream? I never forgot it, but by this time it was very clear that it would never come to pass; I was retired, had no access to a laboratory, and not the faintest idea of how to do it. I actually taught the biology of aging, and all that everyone knew about DNA repair, oxidative stress, the mTOR-FOXO repair systems and all the usual stuff you can find in textbooks and papers. While we knew a lot more about cellular aging, there was still little that could be done. That dream was just a dream, after all.

In general, aging could be summarized as follows: due to damage accumulated over a lifetime, cells develop all of those conditions I outlined before from López-Otín et al.'s *Hallmarks of aging*, and ultimately die when damage overwhelms the cells ability to continue to live; and so, as aging kills cells or makes them senescent, eventually, as it's known, tissues start losing cells as stem cells themselves die and others become senescent. This loss travels up the hierarchy as, if cells are missing or malfunctioning, the tissue they compose no longer functions correctly, and the organs those tissues compose lose functionality, the organ systems of which those organs are a part are then less able to fulfill their functions, and finally the organism functions poorly and this is what we call "aging", and that is what nearly every other scientist in the world and I believed at the time. There was no road to immortality — things wear away, and when the information for their repair (as contained in DNA) is also lost, no hope remains. There is no cellular "resurrection".

While I had plenty of classroom time (online), I still had plenty of free time — and due to the fact that I was teaching the biology of aging I still "kept up" with the field (though it hadn't moved far). I was retired, still in good physical health, an avid hiker (and Utah is one of the best places for that) and was contemplating setting up some online services — my wife and I (she is a computer programmer) were possibly the first people to start an online employment service, which didn't work out, fortunately.

As I was considering my options, I read a magazine piece about the work that was done by Conboy et al.[55] back about four years before (in 2005) that in spite of my fairly broad acquaintance with the literature I had never seen before (and yet it was in Nature!) and I had never heard mentioned, so I sought out that paper and read it with mounting excitement. I now knew how to bring "immortality to humankind" — I was so convinced I told all of my relatives (who generally have a high opinion of me), and friends. But still there I was, a sixty-five-year-old man with no laboratory, and no funding, so though the "impossible dream" now seemed possible, it was still highly improbable, unless I could find a sponsor, and I devoted the next several years — a voice crying out in the wilderness — to find such a sponsor.

What the Conboys' research told us

There is a procedure called *heterochronic parabiosis*, in which "hetero" means "mixed", while "chronic" refers to age. However, it's the "parabiosis" part that needs more explaining, basically meaning "living next to each other", but that parabiosis is considerably more intimate than simply that. In essence, two animals of the same kind, and often genetically isogenous (having approximately the same genes) are severely wounded in such a manner as they can be stitched together. So, in summary, one old animal (typically mice or rats) and one young animal are sewn together. It's a pretty cruel procedure, with a high mortality rate. Often the bigger rat will bite off the head of the smaller one (in rats, there is a significant increase in mass during aging), so why would anyone do such a bizarre and cruel procedure?[56]

The procedure was started in the 19th century; the first works using heterochronic parabiosis (HP) to study aging were done in the 1950s and early 1960s by researchers such as Clive McCay[57] — who had established that caloric restriction extended lifespan — which indicated that if a young and old animal were joined by HP, the older animal appeared younger by virtue of having whiter tendons and other such connective tissue that normally yellows with age. Also, the tissues were softer, as tissue also hardens with age (due to amyloid deposits). At that time, our knowledge of aging biomarkers was limited and few animals had their DNA sequenced.

More to our point, however, it was Ludwig and Elashof who in 1972 determined the effect of this experiment on lifespan.[58] This time, a large study showed an increase in longevity, particularly with paired females, as compared to unpaired animals or same-age parabionts. Later, during the 1980s, there was substantial Russian experimentation with this technique.

The Ludwig and Elashof studies certainly showed an increase in longevity, and (as Michael Conboy first remarked) the younger parabiotic partner was no longer young by the end of the experiments — but what caused this increase? Had I known of these experiments, I would hardly have assumed rejuvenation, but rather that the organs of the younger parabiont compensated for the decreasing functionality of the old partner's organs.

However, I had no knowledge of this HP until I read the Conboy et al. paper coming from Irv Weisman's lab at Stanford.[52] The title tells the story: *Rejuvenation of aged progenitor cells by exposure to a young systemic environment*. The "young systemic environment" was the shared blood supply of parabiotic partners. Progenitor cells? Progenitor cells are stem cells with a limited potency, meaning that unlike, for example, embryonic stem cells (which can differentiate into all tissues of the human body), progenitor cells can only differentiate into limited sorts of cells. So, in the case of liver progenitor cells, there are several types of those in the liver, but only one type that differentiates into hepatocytes, the majority of liver cells. Similarly, there are what are called muscle "satellite cells". These sit on the outside of the bundles of muscle fibers, and when activated, are instrumental in the repair of muscle damage or the growth of new muscle to handle increased physical loads. While referred to also as "stem" cells, as it is clear they only form muscle cells, they must be considered progenitor cells as well. And it was these two types of progenitor cells that Conboy et al. investigated using HP on mice.

There, molecular criteria showed that there was a functional rejuvenation of these types of cells in the older parabiotic partner and apparent aging in the young. As they age, hepatocyte-progenitor cells proliferate at a reduced rate, and the muscle satellite cells are less able to support the formation of new muscle fiber, or to heal wounds in muscle tissue; however, for example, HP returned old muscle satellite cells to near their youthful ability to create new muscle tissue. So now we had proof that the young systemic environment affected progenitor cells at the cellular level. Other experiments (including previous ones) implied to Conboy et al. that changes in inter-cellular signaling were the likely cause of these rejuvenations, but we'll see in a bit that this is not the case. A further and totally convincing study came out of Harvard, and it was one of the most exciting papers concerning the rejuvenation of the old brain by HP, done by Saul Villeda and Tony Wyss-Coray at Harvard.[59]

Saul Villeda et al. summarize their results as the following: "Here, using heterochronic parabiosis we show that blood-borne factors present in the systemic milieu can inhibit or promote adult neurogenesis in an age-dependent fashion in mice."[56] Furthermore, Villeda found that there was an increased concentration of certain chemokines (chemicals that attract white blood cells) in the brain and blood which he suspected were the pro-aging factors in the blood.

In particular, the chemokine CCL11, eotaxin — a chemokine that lures the white blood cells called eosinophils to it — showed substantial increases in concentration, and injection of this chemokine into the tail vein of young mice produced the symptoms of cognitive decline and a dearth of neurogenesis, while HP reversed these conditions, bringing neurogenesis closer to youthful levels. This seemed to clinch the fact that there are pro-aging factors in blood that produced these results and removing them was enough to stimulate rejuvenation by cognitive criteria (spatial learning, pain avoidance) and increased neurogenesis (though this has not been causally tied to learning).

Following this, some of the original authors of the paper, such as Thomas Rando, continued with these studies, ultimately showing that just about every stem and progenitor cell type examined, from cardiomyocytes to oligodendrocytes, was rejuvenated by HP in the older parabiont and aged in the younger partner (though Amy Wagers claims no aging in the young parabionts). Later it was found that simple blood transfusion or even injecting blood serum had similar effects.

The Conboys ran parallel in vitro studies that revealed the same effects of young plasma on old muscle satellite and liver progenitor cells as in vivo.[52] There existed two possibilities for these effects that were not mutually exclusive; either the young blood plasma (cellular involvement was ruled out) contained substances that rejuvenated cells, or the old blood plasma contained factors that aged cells (or both, as mentioned). Further work by the original Stanford group and others illustrated the effect of HP (heterochronic parabiosis), summarized by Saul Villeda, as "In aged animals, exposure to young blood through heterochronic parabiosis improves stem cell function in muscle, liver, spinal cord and the brain and ameliorates cardiac hypertrophy."[60]

My response

After reading the 2005 Conboy et al. paper, suddenly I realized that everything I had been taught (and had myself taught) about aging was false. It quickly became clear to me that cellular aging was a myth. While I worked for many weeks drawing circuit diagrams of redox pathways (I'm more familiar with electrical circuitry, and we are after all dealing with the flow of electrons and their use in producing the energy the cell needs), there was no obvious way that cells could not restore their components given enough energy (in the form of food). Now the reason became clear: cellular aging was a cell non-autonomous process! That means that cellular aging did not depend on the cell's history, but on its environment. Bodily aging was not the result of cellular aging; rather cellular aging was caused by bodily aging (it's a feedback process when aging cells signal organs to change).

It was then that I became convinced that aging could be cured and thought I knew how. Near the end of 2009, during the Christmas season, I wrote to the only person I knew that was brave (or arrogant) enough to dispense with the advice of experts and vow to cure aging through his SENS Research Foundation: Aubrey de Grey. I wrote saying that I believed I had a way of curing aging. I emailed Aubrey, but his gatekeeper, Michael Rae, wouldn't let me speak to him without revealing my secret, and I finally agreed to do so, though SENS Research Foundation would not sign a non-disclosure agreement — but what choice did I have? I knew no one else with the power and influence Aubrey wielded.

My idea was simple — to me obvious. As it was determined that it was the blood's (acellular) plasma that had the rejuvenating effects, why not replace an old person's plasma with a young person's? And the way of doing so was also obvious — a medically proven and certified technique (though used more widely in Europe and Japan) called plasma exchange, a form of plasmapheresis wherein the plasma extracted from old blood would be replaced with the plasma from young blood. My guess was that with a series of such transfusions, whether old blood contained pro-aging factors, or young blood anti-aging factors, or both, what I called Heterochronic Plasma Exchange should do the job.

Well, my revelation did the job, and then the next e-mail I received from the SENS Research Foundation was from Aubrey himself (I have it), and his first statement went something like "Here's why it won't work." Aubrey was of the old school — cellular aging resulted from molecular damage and there's no way young blood plasma could help. He was, however, kind enough to introduce me to the Conboys — but they pretty much agreed with Aubrey, although Irina Conboy suggested that platelet-rich plasma had anti-aging effects. He also connected me with Amy Wagers, who at first sounded interested until she found I had no funding. To the aging community, I'm pretty sure I seemed a crank. However, Aubrey told me that the Conboys would be working with a system together with Frank Longo to try my system, but in spite of a mysterious call by Michael and Irina Conboy, where we set an appointment to discuss the results, they canceled on their end and I never found out the results. The real problem is that in spite of their own results, the Conboys still appeared to support "wear and tear" aging.

Searching the Internet for people who might accept programmed aging, I came across Ted Goldsmith, who, as an MIT-trained systems engineer, could not accept "wear and tear" theories of aging based on engineering principles; there finally I had some company in my beliefs. It was Ted who introduced me to the Russian scientist most responsible for the theory of programmed aging, Vladimir Skulachev. Skulachev, who was the editor of *Biochemistry*, a journal of the Russian Academy of Science, invited me to submit an article to his journal, which I did, called *Studies that shed a new light on aging* in 2013,[61] which fortunately received some interest. But still this was "theory", and I wanted to see results.

It was then that I contacted Mitch Harman, MD, PhD, at the now defunct Kronos Longevity Research Institute, who told me at the time that the institute was funded by the self-made billionaire John Sperling, the man who founded the University of Phoenix as a way to help the working person (such as he had been, a merchant seaman) to get an education. When I told Mitch my idea and sent him my paper, he was very excited to begin with the first recipient of heterochronic plasma exchange. We were able to contact Dobri Kiprov, MD, whose forte was plasma exchange (I found it was surprisingly cheap), and he agreed to do it. Dr. Kiprov had other ideas, however, and was convinced that serum albumin (the major plasma protein) was the secret to rejuvenation, and was hesitant to use young human plasma. I asked that we find young donors, but Dobri, thought "off-the-shelf" plasma was good enough, or better a saline solution of albumin.

Soon afterward, I was invited to John Sperling's home in San Francisco, and met with him, Mitch and Sperling's son-in-law to discuss the procedure. I even agreed to take the treatment together with John, to allay his fears, however that was not to happen. Dr. Sperling, who was 91 at the time, was told by his personal physician that the procedure was dangerous and he recommended highly against it. John was clearly fearful (I could see panic in his eyes) and eventually he declined, and so our big chance was over. John Sperling died the next year and I believe his personal physician did him a great disservice (and me too).

Even a plasma exchange schedule using a solution of physiological saline and albumin would have been enough to promote tissue rejuvenation as the recent work of the Conboys together with Kiprov proved.[14] Apparently, diluting old plasma in half by plasma exchange with saline-albumin produced significant effects after a while, presumably proving that aging was caused by pro-aging factors in the blood (though not disproving anti-aging factors in young blood). Kiprov and eventually Mitch Harman and others I knew formed their own company (Young Blood Institute) but completely abandoned me. Once again, I was nowhere.

On a request from *Current Aging Science*, I wrote another paper, *Towards an evidence based model of aging*,[62] but it didn't bring as much interest as the "Studies" paper did. It seemed that I would never be able to test my theory.

This next part of my journey was rather embarrassing. I was contacted by two people, who I'll simply call Fred and Jean (not their real names), who wanted to sponsor my experiments. Jean, a businesswoman, and Fred, her employee and partner, wanted to start a clinic in Belize, and it would only be a matter of time. Then, their plans changed, and Belize was abandoned for a clinic in the Philippines. After a while, I understood that they were using this as a ruse to attract big money from prominent people without any real possibility of actually setting up a clinic in a nation that was very Catholic and had very limited medical facilities. I cut my contact with them.

During that period, I was contacted by an Indian businessman, Akshay Sanghavi, who was also a very knowledgeable blogger on anti-aging medicine, with a particular knowledge of Ayurvedic medicine — but as I was already committed to Fred and Jean, I had to pass up that opportunity (that is of course before I realized who Fred and Jean were).

Evidence mounts

When a phenomenon that is radically different from expectations — such as "cold fusion" or "polywater" (the idea that water molecules form polymers) — occurs, but year after year results are sometimes negative and sometimes positive, that phenomenon probably has no basis. However, when the evidence for it consistently mounts up, there's good reason to regard it as true.

In confirmation of his earlier (2011) paper, Villeda, in his 2014 paper,[57] successfully repeated the earlier experiments, but added to his prior results an increase in neural plasticity in HP treated animals, measured as an increase in the transcription of genes involved in neural plasticity. Also, there was an increase in the number of dendritic spines on certain cells of the dentate gyrus of the hypothalamus. Similar results in mice were obtained by simple injections of either young plasma (from 3-month-olds) or old plasma (from 18-month-olds), showing that in comparison to old plasma-treated mice, those mice treated with young plasma showed an increase in learning and memory when tested on a water-maze.

Villeda et al. concluded that "Taken together, our data demonstrate that exposure to young blood counteracts aging at the molecular, structural, functional and cognitive levels in the aged hippocampus."[57] However, there was a departure from the earlier paper, which was stated as follows: "Additionally, we previously identified 'pro-aging' factors in young heterochronic parabionts as negative regulators of neurogenesis and cognition; however, these factors were unchanged in aged heterochronic parabionts. Correspondingly, our studies suggest two distinct strategies for reversing aging phenotypes. One possibility is that introducing 'pro-youthful' factors from young blood can reverse age-related impairments in the brain, and a second possibility is that abrogating pro-aging factors from aged blood can counteract such impairments. These two possibilities are not mutually exclusive, warrant further investigation and may each provide a successful strategy to combat the effects of aging."[57]

So, which one was it, "pro-youthful factors", "pro-aging factors", or both? As we'll see, pro-youthful factors will be all that is required — though removing the pro-aging factors would probably speed up rejuvenation, at a cost, but one that may be worth paying.

So, it seemed increasingly clear that my heterochronic plasma exchange (discussed in my 2013 paper[58] as HPE) should work if the same mechanisms of aging and the control of aging existed in humans — and the likelihood that such a fundamental process would differ between mammals seemed to me to be remote. However, how was I to demonstrate this?

Akshay to the rescue

So, it was now about 2016, and it seemed that I would never reach my goal. Well, to imagine that I could rejuvenate people was bizarre — in all of human history, people sought that prize, through black magic, like "the Vampire Queen", Elizabeth Bathory, and with strange chemical concoctions more likely to cause demise than rejuvenation. Besides that, throughout the world there were legends of a magical elixir — the Soma of India, the Ambrosia of the Greek Gods, the Elixir Vitae of the medieval Europeans. The idea that such a possibility existed, much less that I would discover it, was pure fantasy. Yet, the work on HP and similar blood plasma-related experimentation pointed to the possibility.

At this stage, I was feeling my age (74, but I still didn't need two hands to drink), and wondered what I would do with the rest of my life. Not that I didn't have enough to amuse myself — my computer skills coupled with my knowledge of electronics led me to think of various projects, and then, simply grow older, and die. The common fate of humankind; how could I expect better? Yet, that dream I had — the one that told me I would bring immortality to man — never left my mind. I hadn't accomplished my goals, but perhaps others would, though they all seemed to be looking in the wrong direction.

Then I received an e-mail from Akshay Sanghavi, asking whether I would still be interested in doing the experiments I proposed. However, unlike his last offer, made when I was working with Jean and Fred — when he told me I could do the research anywhere I wanted — this time I had to come to work in Mumbai, India, where Akshay (a financier as well as a talented amateur biologist) had started sponsoring projects together with a young Assistant Professor named Kavita Singh at the pharmacology department of the NMIMS University in the western suburbs of Mumbai (don't be fooled, they look nothing like US American suburbs — seeing crows eating a dead rat on the street was not rare — but the rich and famous lived there. That suburb, Juhu, is home to many Bollywood stars.

Honestly, I was very sick (irritable bowel disease) and weak, and the thought of leaving my family for a strange and difficult land to do work I was not sure I could do, and not sure it could succeed, made for quite a decision (incidentally, I feel much younger now than I did then). But what was the point of my life otherwise? Sure, I could rest on my laurels — I had a major input in the discovery of the first breast cancer gene, BRCA1, when I worked at Myriad Genetics in Salt Lake City (but I left with a sour taste in my mouth), I worked for the University of Maryland and was promoted to full professor, and then to Academic Director for Natural Sciences, where I believe I did a good job. I was respected, and while I didn't earn a great deal, I had enough to meet my needs. Impressing others was not any of my objectives, but I did want to contribute to the world, so I applied for an Indian visa.

At the time, the best I could do was to get a two-month so-called E-visa, as all transactions were done over the internet. Eventually everything was ready, and with Akshay's assistance (and payment), I booked my flight to India. Admittedly, after being used by Jean and Fred, I had lost considerable trust in "sponsors". I didn't know Akshay from a hole in the wall, but what choice did I have if I wanted to fulfill my dream?

The flight to Mumbai was to me a horror (stuck in a middle seat for the first 9 ½ hours), then transferring in Hamburg where, because of delays in my flight, I missed my connecting flight and was offered hotel accommodations at a hotel near the airport. This was probably a good thing as I felt at every moment that I was going to pass out. Every movement was painful and exhausting. The next day I took the next 10-hour flight to Mumbai — so I was approximately 20 hours in the air. But eventually I arrived. Then I spent hours on line to have my passport and visa stamped, and another hour looking for my luggage — which in spite of going out of my way to tell the German airport people to make sure my luggage followed me, it did not. I finally got out of the airport and was met by Akshay. I felt that he saw I was too old, but he was committed at that point.

The next day, the airline sent my baggage to the hotel, and I had my first taste of South Indian food for breakfast: idli (a sort of dense bun made of a fermented mix of grains) together with sambar (a spicy red sauce, with no tomatoes), and I was already warming up to India. Akshay was a good host (and since, a good friend and partner), and within the next few days I started on a project Akshay thought promising. Akshay, as I said, was an anti-aging blogger and, understanding many of the symptoms of aging, devised a combination of ingredients designed to mitigate all of them. It was then that I met Kavita, who was our connection to the university, and that was the beginning of our extended family.

The first thing we decided on was using rats rather than mice for our experiment — my original thinking was that the larger veins on rats would allow us to do plasma exchange more easily. Next, we started on one of Akshay's projects. Akshay was also an expert on Ayurvedic medicine, and had his own approach to anti-aging combining both Ayurvedic herbs and other 21st century ideas about aging. The object was to treat the symptoms of aging, all of them with a vegetarian mixture of herbs and oils. My job then was to plan the experiments, and I wanted to cover as many and as varied a mixture of potential markers of aging as we could due to limited money and equipment. Fortunately, as I had never handled animals, a doctoral student — a competent young man, Sagar — was there to help.

The NMIMS University was not designed for research; it was primarily a business school with a separate building in which pharmacology students received their training, mostly to become pharmacists. However, Kavita had an ongoing program for training researchers, both MSc and PhD students. As she was one of the few faculty doing active research, the equipment, mostly for student use, was largely ours, when students were not using it. Unlike in the USA, where each faculty member has their own laboratory equipment and space, at NMIMS all equipment was in common areas.

Later I asked the Dean of the Pharm school, a wonderful woman named Bala Prabhakar (who was voted Dean of the Year, nationwide), why she didn't hire more research faculty, and was told that Indian grants did not include money for the university the researcher worked in, so that the only way they could earn money for the school was to teach — an unfortunate mistake that discourages research. I won't dwell too much on personnel, but Dean Bala made me feel welcome, giving me an office so I could work at the school. I won't forget her kindness.

So, it was up to me to help design the experimental criteria we would use to determine whether anti-aging took place. I wanted to test every aspect of aging, at the biochemical, physiological and cognitive levels. So, by measuring inflammatory cytokines, I took care of the biochemical and physiological levels (plus there were to be many "ex vivo" tests on organs to be done at the end of the experiment). We did have the potential (with begging) of using a Morris water maze (a tank of water in which rats have to learn and remember the location of an underwater platform they could rest on, rather than endlessly swimming), but that was very stressful.

Sagar, who was just completing the final qualifications for his doctorate, built a Barnes maze, which consists of a large, tall, circular table with circular holes along its perimeter — holes big enough for a rat to jump through. Some objects on its surface and on surrounding walls act as visual clues. The table had about a dozen round holes, each about six inches in diameter, and was perhaps four feet off the ground. From all but one of these holes, there was a straight drop to the floor, but one hole had a soft pouch beneath it, one that would hold a rat. In this experiment, a rat was placed in the middle of the table, being fully exposed, a condition that rats hate. However, what they hate even more is jumping four feet to the ground, so the only solution for the rat was to find the hole with the bag, and that's why the visual clues were there, to help the rat orient.

Rats would be given nine days of training, and then allowed to find the hole, singly, and the time taken to find the proper hole (its "latency") was recorded as an indication of learning and memory (also speed and motivation). All-in-all the results were what we expected; the older rats (about 20 months old) took far longer (a longer latency) than young rats (about three months old) to learn.

One indication of aging in all vertebrates is an increase in inflammation — and this could be measured by the levels of pro-inflammatory chemicals, made by the body, called "cytokines" (these are signaling molecules found in everyone's blood to varying degrees). The cytokines I chose were the most important with regards to inflammation (only two were chosen — though I would have liked to include IL-1beta — simply for lack of resources and personal).

Sagar had to force-feed Akshay's mix to the rats via syringe as they wouldn't eat it on their own, and eventually after a month of testing, nothing. We were going to sacrifice the rats at the end of the month, but I thought it was worth an extra month — as most Ayurvedic medicines take some time to take effect — and by the end of the next month, the rats did appear younger. In fact, their inflammation returned to near youthful levels and their maze-solving abilities increased as well. However, translated into human terms, a month of a rat's life is worth about 2.5 human years (as a ratio of average lifespans), so that someone would have to eat fairly large amounts of this concoction for five years to see effects (and if rats won't eat it…).

While this is certainly better than the alternative (growing older during those same years), another factor made us abandon our efforts to publish these results; the old treated rats had lost considerable weight when compared to untreated old controls, making it possible that the well-known phenomenon of caloric restriction was responsible for the apparent rejuvenation. Though we weighed the food eaten, we did not weigh the feces. It is still possible that Akshay's formula does work the way he expected, and it would be wonderful to have a plant-based (mostly) anti-aging medicine, but our later results pointed in another more promising direction.

Don't feel bad for Akshay though; at that time, we implemented another of Akshay's patented ideas, an anti-aging compound in an easily deliverable form which I used myself (as it was generally regarded as safe) with excellent effect — it gets rid of the horrible "actinic purpura" (purple spots caused by sun damage during youth) as well as, surprisingly enough, improved both my coordination and my energy levels.

But now it was time for my project; the idea, as before, was to do a plasma exchange with rats. To our knowledge there was only one group in Germany that had done plasmapheresis on rats, and they promised to send us the information we needed. However, in spite of constant inquiry, they never did. So, I was completely stuck, with nothing to do and little or no hope of doing what I wanted to do. I had an idea: if we could not do our experiment with the German apparatus, perhaps we could do it manually.

Plasmapheresis is basically this: blood is withdrawn from the subject, and centrifuged to separate the cellular part (about half of the blood volume) from the plasma. The packed cells, mostly red blood cells and platelets, form a solid red mass at the bottom of the centrifuge tube with a thin layer of white blood cells above it (the so-called "buffy coat" layer). The yellowish (or reddish, if you have hemolysis — breaking of red blood cells) plasma is then removed, and an equal amount of a saline solution with added serum albumin (to maintain osmolarity) is then mixed with the cellular layer and is re-injected into the patient. The main purpose of this procedure is therapeutic — for example, rapidly (though not so rapidly) removing toxic substances from the blood plasma. However, if you've ever been given plasma (often donated by young people for money), the process is identical except for the amounts of plasma taken — ideally, in therapeutic plasma exchange, nearly all plasma would be replaced. The procedure (plasma exchange) is medically approved.

A rat has approximately 30 mL of blood (hence about 15 mL of plasma). If we could perhaps withdraw five aliquots of blood, centrifuge them, remove the old plasma and replace it with equal volumes of young plasma, mix and re-inject it into the rat, we could get the same effect as with the automated plasma exchange normally done on humans (with a single volume of blood — 30 mL for a rat, 5,000 mL for a human). However, this too was not to be — the rats' veins were too fragile for this sort of operation, so again I was stymied. Again, no hope of doing what I needed to do, so I decided to do something else.

By this time, it was my third two-month stay in Mumbai, and I actually had to wait a considerable time before I could apply for another (the number of visits per year was limited). So, I had to come up with something new, and quickly. At approximately the same time, Sagar had completed all his requirements for a doctorate, and was leaving to take a post-doctoral position with a large pharmaceutical company in the USA. So, we needed to find a replacement for him, someone who could work with animals, and with an interest in molecular biology. We did have a graduate student at one time, but her work was subpar and anyway, she left us for greener pastures (had she but known).

We placed ads and conducted several interviews but no one seemed to have the skill and interests we (at this point, Akshay, Kavita and I) needed. That was until Shraddha Khanair. Truth to tell she was a small-town girl, and probably mostly spoke Mahrati at home, so I could barely understand her English, but if I strained and asked her to repeat things I gradually came to know that she had a good grasp of molecular biology and could handle animals, and I voted yes on her, and she became part of the team. Now we were all inextricably linked like a family should be, all out to help each other.

Both Kavita and Shraddha are Kshatriya (the warrior and royal caste). Shraddha is a direct descendent of Shivaji Maharaj, who established the state of Maharashtra in which Mumbai is found. Akshay, the businessman, is a Vaishya, the class of businessmen (truthfully, he's a lot more than a businessman). I myself am a US American Jew.

It's funny, I never thought of the caste system this way, but if none strives, there is no competition; life, theoretically, becomes easier for the higher, "noble" castes, at least where no one seeks to rise above their "station". In that system, even the manual extractors of manure from the often clogged, open sewers of Mumbai knew it was their job and their fate (presumably for something they did or didn't do in a prior existence), and their job was their religious duty. Your lot in life was determined by your family's heritage, its caste.

A US American Jew has no caste, except among Indians, rich ones — I met several who were Jains (monotheists, but with a formless God who is everywhere, not an old man with a beard, and a desire not to hurt any living thing), and discussed at the table the business acuity of Marwaris (a Rajasthani group), Gujaratis (Akshay is a "Gugu" as my friend Tina calls them — her ex is one, though) or Jews.

The recognition of differences between peoples in a land with 122 official languages (and many more dialects) is very important. You should also keep in mind that India is 1/3 the size of Europe with twice the population.

The different, not readily accepted peoples were the Muslims, and that's the result of the Moghul invasion and conquest of India for many centuries. So, these people (some of whom have been very successful in India and elsewhere) don't have a caste (though you might assume they were lower caste with aspirations for betterment), so they don't seem to be part of the people and represent an ancient humiliation (and well-known to everyone).

For Muslims, Hindus are the worst of idolators, people who Muslims could lawfully kill if they would not convert to al-Islam (the surrender). Conflicts between Hindus and Muslims are frequent and deadly; of course, Hindus are the majority. Numerous other religions and peoples were received by India — oddly enough Jews had a very long history in India (the Sassoon family are Indian-derived Jews), and also the Parsis (Zoroastrians who escaped the Muslim conquest of Iran), and many, many other peoples. With a short drive out of the city (not that any drive to leave Mumbai is short) you can enter tribal territories where the people still live as subsistence farmers.

8

The invention and discovery of E5

Though I was sure that my procedure would work, there were insoluble problems for us. Was there a different approach that could give me the same effects? My original thought (expressed in my 2013 and 2015 papers) was that there were pro-aging and anti-aging factors present in the blood — but which mattered more? While only in 2020 the Conboys et al. demonstrated that diluting old blood plasma was sufficient to produce an appearance of rejuvenation at the tissue level,[63] experiments such as those done by Wyss-Coray injecting plasma (even human plasma) into mice already showed rejuvenation.[57,64]

However, as these injections were always considerably smaller than the blood content of a mouse (75 μL compared to around 1500 μL, so five percent of blood volume per injection) and by the time of the next injection many plasma proteins would have been replaced (they were given on a biweekly basis over months), there was no time at which one could say that dilution of plasma was the cause of this rejuvenation, as plasma is a dynamic liquid ("You can't step into the same river twice"). It also occurred to me, as the experiment was stretched over months, that these youth-inducing factors had a long half-life or else their effects wouldn't have accumulated over time.

So, while it was still clear to me that there were pro-aging substances in the blood, it was equally clear that there were pro-"youthening" (as we shall see) factors as well, as simply injecting young plasma into mouse brains was sufficient to cause a return of neural plasticity (assessed at the cellular level by increased production of the mRNAs that are specific for neural plasticity) and there were indications of increased memory and learning from slices of hippocampi that showed increased *long-term potentiation* (LPT) which is associated with learning and memory.[62]

The paired hippocampi (right and left) is one of the sites where neurogenesis occurs. In humans, the lining (the periventricular space) of the internal cerebrospinal-fluid-filled ventricles at the center of the brain is another site; a third site (though perhaps not in humans) is the olfactory bulb, a primitive brain region concerned with the sense of smell. Young blood also increased the density of neural spines, tiny projections that are produced at every connection (synapse) with other neurons. It was also clear that these effects could be overpowered by factors present in old blood. At this point, I believe I understand why, and will explain later when discussing our results.

So, I read as much as I could, and found that projects that failed to show rejuvenation, though nearly identical to Wyss-Coray's serum injection experiments, had more to teach me than the ones that succeeded. I also, by this time, knew what to look for and where to look. At the end, I knew what not to do, and what to do. But that was in my head; the question was if it would work, and time was growing short and we had to try.

Mumbai is not convenient, as traffic is horrible, and the city is widely scattered, so it took a long time to gather what was needed (and we had to import old rats from a distributer hundreds of miles away, meaning air transport). The roads are something else — it took us 12 hours to go 300 km to Shirpur (where the second NMIMS campus is located) to guest lecture the students; however, Shirpur is rural India, and that is a whole other book.

Nevertheless, ultimately, we were ready to go and I was ready to leave. We prepared our first batch of E5 — and we were going for broke. Akshay said we should increase the dose to work against the remaining pro-aging factors, and I agreed, so each old experimental animal was given a dose by Shraddha (I helped, but mostly watched, as she's very good) that was a multiple of the amount of E5 that we believed a young animal would have (evidently, we had to make many assumptions). So after Shraddha (who had helped me in every way, even helping order lunch from the school cafeteria, in Hindi — the food was South Indian, delicious and cheap) gave the six rats four injections of less than a milliliter in the rats' tail veins, four times, on alternate days, I packed, and took another excruciating (for me, I know some who love it) twenty-hours flight (plus lots of waiting).

I went back to my family in Salt Lake City, and quite honestly I was a bit relieved. If it worked, even slightly, we'd work on perfecting it, but unless a whole bunch of sequential guesses was correct, it wouldn't work at all, and I could remain with my family. I had done all I could do. If I was wrong, I wouldn't be the first. Like any businessman (and I was only just beginning to really trust Akshay after our first real argument — distrust being a residue from past acts of betrayal), Akshay would have to eat up the losses. I hated that idea, but when you take chances, it's to be expected. Truthfully, I didn't expect it to work, but it was the best I could do. Yes, it was crazy — I felt like I had been given a mission, but I had done my very best, so whatever happened, at least I knew I did my part.

It wasn't a week until I received an email from Kavita (Shraddha is her grad student) saying that the grip strength of the old experimental rats had shifted considerably, much more than I'd imagined. And soon, the levels of the cytokines we measured (Il-6 and TNF-alpha) started declining towards youthful levels. And it didn't take statistics to evaluate our results, although Agnivesh, Kavita Singh's husband, a pharmacologist, and an extremely nice person, did the statistics for us. The results were clear to the naked eye.

Well, I'll admit that I had no choice, and by that time I got a five-year visa (with the help of Akshay's influential friends), with unlimited returns and departures. I packed my bags again, said goodbye to my wonderful and supportive family, and took off to India again. This time, I believe it was in the Netherlands that I waited for the Indian flight, but quite honestly, I enjoyed it a bit. I was truly rejuvenated by getting up and doing something worthwhile. Apparently, my string of somewhat strange guesses had proven correct. It was probably a good thing that I didn't know Big Pharma spent millions trying to track down an E5-equivalent and failed. Anyway, it was a shot in the dark based on a limited understanding, but it proved better than understanding incorrectly, if I was right.

I came back to India, but the Airbnb I had with Milan (a guy's name, it means "meeting") was taken, so after searching, I came on an address in

Juhu Beach, maybe a block from the ocean (Indian ocean — people swim there, despite the raw sewage emptying into that water maybe a mile south).

Anyway, the landlady (the Airbnb owner) was Tina Pandey, and she said she'd be out of the apartment most of the time, so I'd have it to myself. That sounded good — yet Tina was interesting to talk to, very interesting. Finally, Tina left to go to another city where she had family, so I finally had the apartment to myself. When Tina returned, we both realized we missed each other and we have become the best of friends. Tina was an imposing woman, though considerably less than five feet tall (Indians are tall — I'm 5'9" and I'm short in India, at least in the higher castes) and weighing in at 70-something pounds. Describing her would really require another book.

Life in Juhu had me exercise along with hundreds of Indians who walked through the sand of the beach (you could occasionally find very strange-looking dead fish being eaten by the ever-present crows — not totally black but with a dusty-looking head).

At my passage from street to beach were set up a number of religious shrines to various Gods and a huge "RAM" (as English-speakers call Lord Rama). India is super-religious, with dozens of religions and a greater variety of people and costumes than you can imagine. The remains of the flowers used in a Friday pooja (prayer ceremony) are all over the beach as they have to cast them into the water, and it's crowded — but beach running (more like walking in my case) is good exercise.

Back in the lab, we readied ourselves to try the same experiment again, and again much time was spent in getting basic necessities, but we performed the experiment again — there were eight young control rats, eight old control rats, and eight old rats of the same age (around 20-months-old) that were treated with E5.

Again, the same schedule was assumed, measures of inflammatory cytokines were taken, and in what subsequently became our assay to the effectiveness of our treatment, cytokine levels started to fall on the fifth day following the first injection.

At that time though I had an idea; we had over thirty biochemical, physiological, and cognitive age-markers we (mostly Shraddha and Shivani, a technically very competent young woman) tested for. Several were tested (for organ levels of important chemicals) ex vivo (sacrificed). At the end of thirty days, we had an excellent repetition of our first results — it wasn't a fluke. All of the more than 30 bio-age markers, from biochemical to higher-level cognition, showed a significant reduction in age — but not any of them were definitively related to age, so my idea was to contact the world-famous scientist who was the force behind epigenetic aging, Steve Horvath, the man best known for the DNA methylation clock he invented that could estimate a person's age within three years by DNA alone. I proposed that we work together on building a rat clock (for Sprague Dawley rats), although there was no certainty, at that time, that such a clock was even possible. That required dissecting dozens of animals of all ages, as organs had to be dissected out and their DNA separated and purified. The animals were sacrificed in six intervals from birth to old age. The explicit proviso was that Steve would process our treated (and control) animals to add the most definitive age testing, the "gold standard". To my surprise, Steve agreed.

To be honest, I had no idea whether Steve's tests would show anything — at least towards the end, when I had figured out alternative (wrong) pathways for how this apparent rejuvenation might occur — but if there could be a clock for rats as well as humans, that would go a long way to support Steve Horvath's theory of epigenetic aging. But before I go on, I'll have to tell you a bit about the subject of DNA methylation and Horvath's clock — now clocks. Let me first tell you what I'm talking about.

DNA methylation

At this time, everyone has some idea of what DNA is and what it does. A brief summary is given in defining two words:
1. Transcription: single-stranded RNA copies are made from DNA, which encodes instructions on how to build protein molecules in a "genetic code", translated into a protein strand by the ribosome, a molecular machine. The laws of base pairing apply to RNA and DNA, which differ in a single hydrogen atom in the ribose (a five-sided ring) sugars to which the four sorts of complex, nitrogen-containing "bases" are attached to make a nucleoside and ultimately a nucleoside triphosphate (the triphosphate stores the energy used to add another nucleotide to the growing chain of RNA or DNA). While DNA uses the base thymine, RNA uses the base uracil, though both bases have the same binding property — they bind to the base adenine if it opposes them on the other

complementary strand of double-stranded nucleic acid polymers. RNA is usually single-strand, while DNA is double-stranded.

A single strand of a nucleic acid might be represented as such: pApTpCpGpApGpApApApCp... Where A, T, C and G represent the nucleosides adenine, thymine, cytosine and guanine, and the "p" represents the phosphate bonds that bind them together — a polymer is similar chemicals groups joined together, head to tail (as it were) to make a long chain. The single-stranded DNA of a human has about three billion (3,000,000,000) of these nucleotides strung together, and these (among other important things) carry the codes for making proteins; that code is copied onto RNA and transferred out of the cell's nucleus (where its DNA is kept) into the cytoplasm where translation occurs.

Protein molecules called "transcription factors" decide which parts of the very long strings of DNA are transcribed, by fastening to specific sites on DNA; often several different transcription factors attach to multiple regions near where transcription of DNA begins, in "promotor" regions of DNA, which are non-coding, regulatory sequences.

2. Translation: an amazing molecular tool, the ribosome, similar in structure and operation from bacteria to humans, "reads" the RNA — called messenger RNA (mRNA) because it carries the message of which proteins the ribosome will build, by providing the protein-specific instructions for doing so.

MicroRNA may decide whether an mRNA ever leaves the nucleus to be translated to protein or whether it will be destroyed in the nucleus by a macromolecular complex designed to slice and dice it.

Now, take a look at the original short polynucleotide I wrote above:

pApTpCpGpApGpApApApCp...

By the way, the double-stranded version is:

TpApGpCpTpCpTpTpTpCp...

pApTpCpGpApGpApApApGp...

RNA is usually "single-stranded", but can form many different shapes with self-pairing — part of any RNA chain is complementary to some other part of itself, and so the RNA can loop around and bind to itself to form a potential infinity of shapes. As it is said by biologists, form follows function (and vice versa), and so these complex molecules can do more than simply carry information, like the conformationally static DNA.

RNA molecules can do things on their own, as proteins can. Ribosomes without any of their 80 or so proteins (in mammals), just the three intricately folded, long RNA molecules that make up the heart of the ribosome, can still assemble peptides from mRNAs (given the raw materials they need). RNAs have been found which are self-splicing (first in the ciliate tetrahymena), and those are a vital component of telomerase, as the TERC gene codes for the RNA that is the template the telomerase reverse transcriptase uses to elongate telomeres.

Also, if you were to partially cleave that DNA sequence with enzymes called "exonucleases" that break the phosphate bonds holding the chain together, you'd get Ap, Tp, Cp, Gp, and maybe some dinucleotides like ApC or GpT and longer chains as well, right? Considering that all these four sorts of nucleotides are present in known proportions, you'd think that the dimer ApT should appear, by pure chance, as often as TpA, and you'd be right, and you'd think CpG should appear as often as GpC, and you'd be wrong.

The dinucleotide CpG is much rarer than it ought to be, statistically, apparently because it has a special function. While instances of this CpG are scattered throughout the genome (the sum total of an organism's genes) they are particularly concentrated in the control regions of important sets of genes; these CpG rich regions are called "CpG islands". These are simply stretches of about 1,000 nucleotides in which the ratio of CpG to GpC is higher than 0.6 — but the real importance of these islands is that they lie at the promoters of genes and their message is encoded in the first part of the "transcript" (mRNA). To make it clearer, CpG sequences are preferentially found in controlling regions of the DNA. It is also possible to add a very simple chemical "group" (an arrangement of atoms that stick together) — a central carbon atom with three hydrogen atoms sharing electrons around it and an unattached bond sticking out, called a methyl group (Figure 20).

And there are enzymes that attach methyl groups to cystine (C) residues of CpG groups, and oftentimes when the CpG group in these gene-controlling CpG islands is methylated (has a methyl group, or hypermethylated if there are several methyl groups on it), there is a change in the output of a gene, usually negative. So, methylating a CpG group negatively affects the transcription of the gene it "promotes".

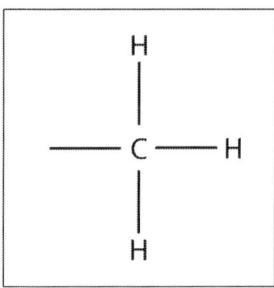

Figure 20: Methyl group.

The interesting thing is that some genes are downregulated with aging, and some are upregulated with aging (especially inflammation). Typically, CpG islands become hypermethylated with time, and those methylated regions outside of the "islands" become less methylated (hypomethylated). Some genes change their methylation status with respect to time in a way that reveals the age of an organism, and many believe, as more evidence is discovered, that these methylation patterns produce the age-phenotype of the organism.

By using DNA sequencing techniques that can distinguish between "Cs" and "methylated Cs", and using statistical techniques together with artificial intelligence that can spot patterns in volumes of data too large for our limited brains to store (much less be able to assess the interconnections between all those data), Dr. Horvath (people call him Steve, he's universally liked, I would say) came up with a limited number of DNA CpG sites where the percentage of these sites that were methylated was the primary input data to an AI which compared this percentage to many ages and organ defined age patterns of DNA methylation (DNAm).

Steve believes, and I too believe, that aging is an epigenetic phenomenon; just as the epigenome determines cell differentiation, the differentiated state of a cell is determined by several epigenetic means. And let me clarify: epigenetic phenomena control the expression of genes and even the form of the gene-products (usually, but not always, proteins) without changing the sequence of nucleotides that constitute the genomic (and mitochondrial) DNA. Every cell in our body has the same exact set of instructions, yet some become liver cells and some become neurons. It is our presumption that cellular aging changes, like those that cause cell differentiation, are controlled by epigenetic means. It remains an open question as to whether or not the changes in DNA methylation are the cause of aging or only yet another output of an "aging clock", the cumulative result of oxidation damage, protein aggregation, telomere attrition, and metabolism in general, as it is known that metabolic enzymes are found in the nucleus (some serving as both enzymes and transcription factors — they're one molecule or one molecular complex with several functions, sometimes completely unrelated, it seems).

During development, as it is usually conceived, from zygote (fertilized egg) to adult, there are a number of named life stages, e.g., embryo, fetus, neonate, baby, toddler, school child, tween, teen and adult, and for biologists, at least those concerned with aging, that is the end of development — all the rest is deterioration. However, in the world of the social sciences, there are post-adult developmental stages, each with its goals and characteristics, and the social, but not biological scientists, understand that these post-adult life-stages follow a stereotypical progression extending for a period — latter life-stages are those with increasing hazard rates that eventually lead to aging disease and death. Steve Horvath has shown, both in humans and rats (and more recently in all mammals[65]), that aging can be followed with the same clock. Feed into Horvath's algorithm to what extent homologous CpG sites are methylated and you'll know you have either a 1 ½ -year-old mouse or a 36-year-old human.

What we must make of this is a lot like the basic theory of Neill that a predetermined passage through preassigned developmental phases leads to death. The Horvath clock recognizes these developmental stages (as they are characterized by changes in DNA methylation), such that a middle-aged mouse and a middle-aged human have the same patterns. It is obvious that lifespan is a species trait that depends on several factors but mainly the species' ecological role and its need to survive as a species in Nature, which is largely controlled by predation, as Robert Ricklefs asserts. He also asserts the reality of the fixed relationship between immature and sexually mature periods in all terrestrial vertebrates (the relation depending on class [bird, reptile, mammal, amphibian]), and acknowledges that the length of immature development is related to predation, as is post-mature lifespan, but admits he doesn't know how this relation works.

Thus, the underlying idea is a clock based on the periodic changes in cellular redox potential occurring on a daily basis, and the cells' cytosol becoming significantly more oxidizing with age (or rather advanced life-stages); maybe this is the molecular basis for why some enzymes might work in youth and be absent or work differently with aging (the "redox code"), as redox changes occur in the cell. But as they say, "the proof is in the pudding".

Back to E5

We were all pretty excited about our first month's results — we had shown our original experiment wasn't a fluke. This was not surprising, with all eight rats we treated showing 100% efficacy; there was no chance it was a fluke, and after about 90 days (where there was a rise in inflammatory cytokine levels to about half of those of old rats), we were going to sacrifice our rats to assess their organ levels of various age-related biomarkers (which I'll show you later), when I thought "No!". For what we thought was the sake of publications (and getting investors), I said "let's sacrifice two rats of each of the three groups (controls — young and old — and experimental) of eight animals and give the remaining six experimental rats another E5 treatment (four injections in the tail vein) to see what happens". My big fear was that once the body was fooled into believing it was young, it might develop "defenses" (in its suicidal endeavors) against E5. However, we found nothing like that occurred; in fact, the opposite happened, as I'll show you.

Well, we had what we wanted, more than 30 different biomarkers proved that we had significantly reduced the age of old rats. What more could we want? I left my friends in Mumbai, the Nugenics Research family, and my friend Tina (who brought me to the airport — she was no longer my landlady, but my friend). I returned to the United States to resume life, hoping that Akshay would be able to get enough money for us to confirm our discoveries with third-party validations and extend E5 experiments, bring it to market and... Let's be real: change the world.

However, the most exciting news was totally unexpected, when we received an e-mail from Steve Horvath telling us our experiment worked, and worked impossibly well — the rejuvenation indicated by all our physiological and biochemical data, which seemed to show our old rats to half their chronological age, was confirmed by Steve Horvath's new rat clock. Their DNAm age was actually less than half of their chronological age.

2020 was possibly the worst year for so many things, and Akshay was running more than five businesses, paying for all of this and meeting with us periodically as his schedule would allow (we were clearly his favorite). He is knowledgeable and intelligent (as well as "thrifty" — I'll admit to the same), but generous too. We are now incorporated under the name Yuvan Research, producing E5 for third-party validation by a Contract Research Lab (CRL) and academics, including Steve Horvath's and Greg Fahy's groups.

Show and tell

Some (most) of these results were published in our preprint on bioRxiv,[1] but the paper has not been published, because we will not settle for lesser journals. And top journals really can't publish our work until we reveal what E5 is. Fair enough; scientific experiments must be repeatable to confirm results, and without available E5, there is no way to confirm them. However, that doesn't mean we might be making everything up out of our heads. bioRxiv reviewed all of our primary data, and found it on the "up and up" and we know it's true.

A lot of the following stuff would be pretty boring in a scientific paper, but since it promises you, at the least, a second life, that makes it kind of pertinent to everyone. So I will now present our data and talk about what it means. Each graph really has a story behind it; some you know already, some you don't (unless you are also a student of aging — post-adult development). But these are the most exciting results anyone has ever seen.

Object #1

The first object I'm going to start with is our first indication that our preparation, which we call E5, worked.

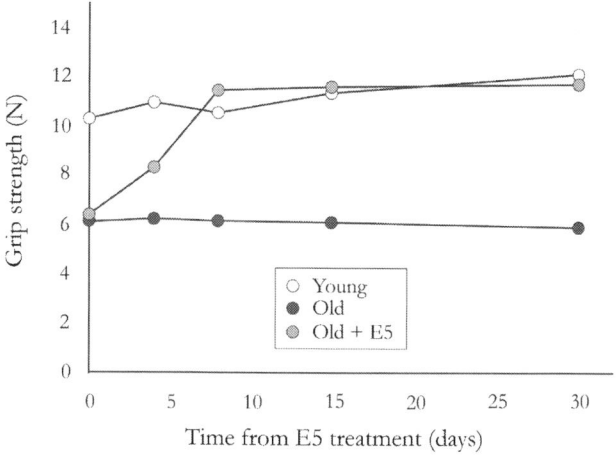

Figure 21: Grip strength variation due to treatment with E5.

The results shown in Figure 21 are from our first experiment, the one I thought had little chance, but happily I was wrong. As you can see, the grip strength of the young three-month-old mice remained high ("N" stands for Newtons, a unit of force, equal to about a quarter of a pound), increasing a bit with age (the white circles). The older rats (untreated) showed a slight but steady decline in their strength even over 30 days (which, remember, is 2 ½ "human years" for them).

But, of course, the exciting part is the light gray circles of the treated old animals. The animals from our old control group (dark gray circles) were of the same age and sex (male) as the group we were treating (the experimental group). The experimental group animals (light gray circles) were equally weak starting out on the day of their first injection, day zero (the young controls were three-month-old rats), but by five days after the first injection, the experimental group was already intermediate in strength between our young and old controls. By the tenth day (the days are on the horizontal axis) they actually over-shot the young group in strength. At the end of the first 30 days (31, actually), we sacrificed the rats (nitrogen, an easy death; I almost died that way so I know it from personal experience). Their organs were sent to independent commercial labs for pathology, photomicroscopy, and then were assayed for the levels of various biochemical markers of age.

What does grip strength mean and how do you measure it? The apparatus for measuring grip strength has a meter at one end with a rod attached to it; when that rod is pulled, the meter measures the force of that pull. At the other end of the projecting rod, it has a series of bars a rat can grab onto (as shown in Figure 22). One student holds a rat (the big ones weigh more than a pound) by the tail and allows it to grab the bars of the grip-strength meter, and then the rat-holding student (or person — it has been me) pulls the rat's tail until the rat is forced to let go of the bars and continue on its downward path (as the test is performed from a small height and rats are afraid of heights), and the grip strength meter records and displays the maximum tension.

We didn't repeat this in our next set of experiments because I thought, as did Shraddha, that there was too much of a human factor present, and more importantly, the person who loaned us the meter wanted it back, and it was too pricey for us to buy one at the time.

So, what are the implications of this assay?

As I see it: the change in strength undoubtedly occurred before any cell division took place, so at least some of the change in strength with aging is not due to loss of muscle mass (as the muscle satellite cells that replace injured muscle don't respond to signaling — Notch signaling, according to the Conboy et al. paper[66] — or do so by making non-contractile fibers, not useful muscle). Muscle mass does decrease with age, but we did not observe muscle mass increase, as we did not weigh individual muscles; however, we did show that there was an increase in muscular strength when the old animal was given a youthful internal environment that resets the old muscles to a younger age-phenotype.

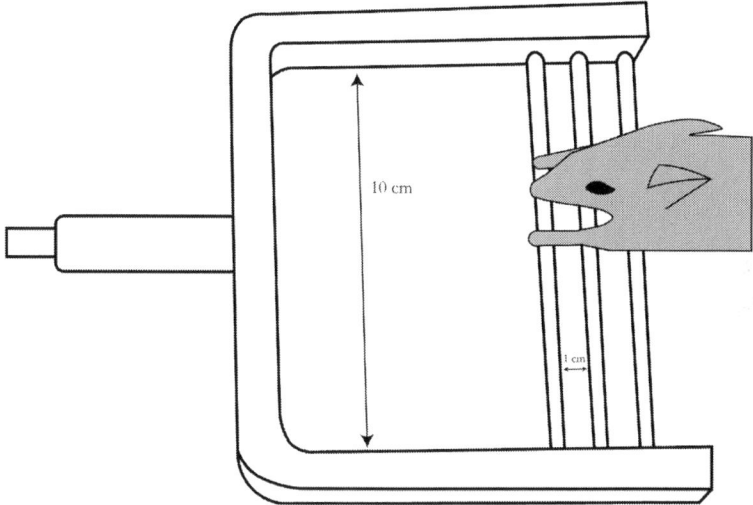

Figure 22: Illustration of the apparatus for measuring grip strength.

I don't know if rats "bulk up" to impress other rats; I suspect they do not (we worked with males mostly), and we don't know if myogenesis (the formation of new muscle tissue) takes place, but there seems to be a response at the cellular level that makes the muscles stronger. As frailty is an inescapable and untreatable part of the aging process, we believe E5 to be an important solution to this impossible-to-solve problem. We have already purchased new grip strength meters, and we continue to work with rats in our Mumbai laboratories as there are many, many questions we want answers to (and rats are convenient, compared to dogs or lawyers). We are also treating rats with E5 "forever", should they live that long, and seeing if grip strength remains constant in those treated rats. It may be possible to bring cells to the age-phenotype they would have had at early life stages, perhaps allowing us to regrow organs. We believe we have uncovered a new continent and are just seeing its highest peak; I believe what lies beneath will offer opportunities that presently lie beyond our imagining.

The continuous and life-long (however long that might be) administration of E5 is an experiment generously funded by Didier Cournelle,[67] whose interest is life-extension, and whose question is: how long can we extend rat life? He bets we cannot exceed maximal lifespan (about four years) by 50% — I'm quite sure that's one bet he wants to lose. Experiments with non-human primates — monkeys — will bring E5 closer to human use, and using E5 on dogs, apart from validation, will show how we might rejuvenate our old companion animals. And wouldn't that be great?

Object #2

The two graphs in Figure 23 represent the concentration of two important inflammatory cytokines (those internally produced signaling chemicals responsible for the chronic inflammation associated with animal aging from fish to man) over the course of the two-treatment experiment I mentioned earlier (our second set of experiments). Again, dark gray (the first bar in each set of three) represents the old controls, the light gray bars (middle) the experimental group, and the white, our young controls (although at eight months old they're not so young anymore). The down-pointing arrow above day 95 of the experiment (actually day 96) shows when the first injection of the second treatment was given (the first of four injections, one every other day — so about a week of injections).

Now, let's look a bit more closely. Notice the top graph shows the blood levels of the inflammatory cytokine Il-6, and the bottom graph, the levels of the inflammatory cytokine TNF-alpha (tumor necrosis factor-alpha). Since their levels vary similarly, let's look at TNF (the "alpha" is now "politically incorrect"). As expected, the blood concentration of TNF starts out the same in both old control and experimental groups (of the same age), the levels in both being about three times higher than in the young controls. However, even four days following the first injection of E5, TNF levels went down 15% to 20% (in all animals). By day eight, they had reached their nadir and started to climb. By day 95, the day of their first injection of the second treatment, the levels of TNF in the experimental group had climbed to near midway between young and old. We can see that although aging seemed to be rapidly reversed (with respect to inflammatory cytokines), aging, in terms of the comparatively more rapid increase in TNF levels in the experimental animals following treatment, appeared to proceed more quickly than normal, but at the end of what for humans would be eight years (2.5 human years/rat month), the treated old rats were still "younger" (in terms of inflammation) than the untreated animals.

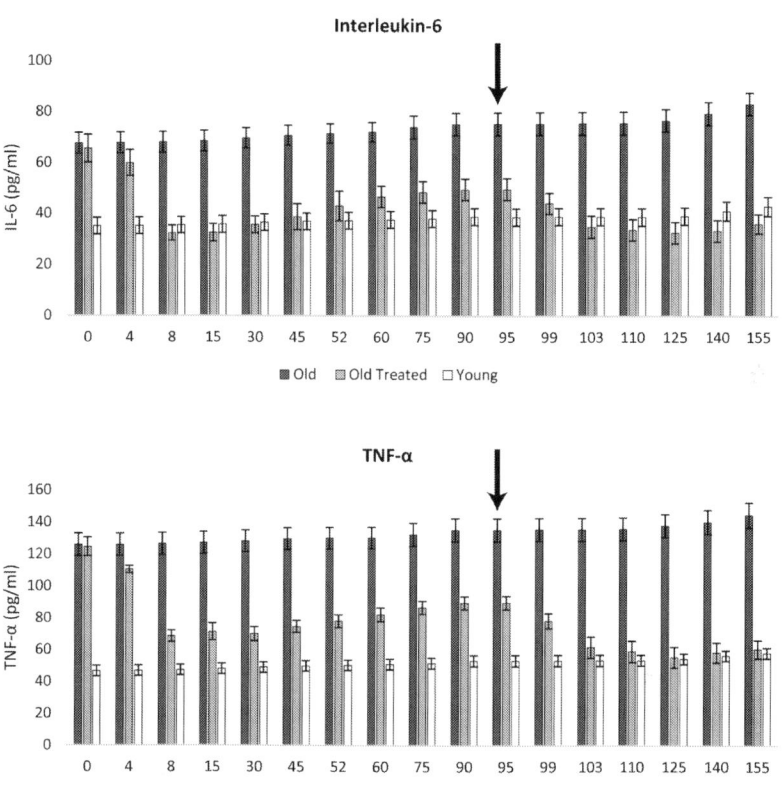

Figure 23: Variation of the concentration of IL-6 and TNF-alpha in the treated and control groups. Adapted from Steve Horvath et al.[1], CC BY-ND 4.0 https://creativecommons.org/licenses/by-nd/4.0/.

So, now we look to the right of that purple down-arrow indicating the beginning of an identical second E5 treatment. Here we see that the TNF levels go down to their lowest point just four days after the first injection of the second set, and they keep going down to points lower than in the young controls and stays at that level even 60 days (the rat equivalent of five years) following the second E5 treatment. It seems that after a second E5 treatment, aging goes forward at normal or even reduced rates. The COVID-19 pandemic stopped this experiment (as our facility at NMIMS University was shut down), but as mentioned, a new and hopefully very much expanded (in terms of E5 treatments) experiment has begun.

The situation is similar in the Il-6 study above. Again, it takes four days to see a fall in its blood levels, but by about day eight, the level of Il-6 is now below (slightly) the levels in the young controls! After the second treatment, the levels of Il-6 (for which the notorious transcription factor NF-kB is responsible) went down again. NF-kB is elicited by high levels of ROS, and NF-kB binding to DNA is responsible for much of the harm that senescent cells create, including their production of inflammatory cytokines, and further oxidative stress makes cells "possessed" by NF-kB immune to self-killing (apoptosis — a mechanism for getting rid of cells that "go bad"). We see that the levels of Il-6 in the treated rats are below those of the young controls (which began at three months old and were eight months old by the end of the experiment), and do not seem to rise above them even 60 days following the first injection of the second treatment.

Note also that in both young and old controls there is a steady increase in the levels of both Il-6 and TNF with age.

What's the significance of this? First, the theoretical: we know that all vertebrates (terrestrial at least) have increasingly higher levels of inflammation as they (we) age, and many explanations have been given for this, including unresolved infections, and viruses hidden in the genome unraveling themselves from the host's DNA; however, with the evidence we already have, we can definitively state that the chronic inflammation of aging (*inflammaging*), is caused by the lack of E5! Our results demonstrate the natural rise of these cytokines with age, and that resetting the animal's age resets the levels of inflammation.

Practically speaking, chronic inflammation is believed to be the cause of several diseases, including cancer, heart disease and dementia. By eliminating the chronic inflammation associated with aging, we should see a marked decrease in the appearance of the many "diseases of aging" associated with them. The increase in grip strength is an excellent sign that E5 might treat frailty. But is this the beginning of a pattern? What pattern? Perhaps that age-related changes have no cause other than age. This is what Stroustrup's formula, $r(t) = t/\lambda$, really tells us, that there is no cause of aging mortality other than an organism's passage through their latter life-stages — everything else follows.

Object #3

Next, let's have a look at several other biomarkers in Figure 24.

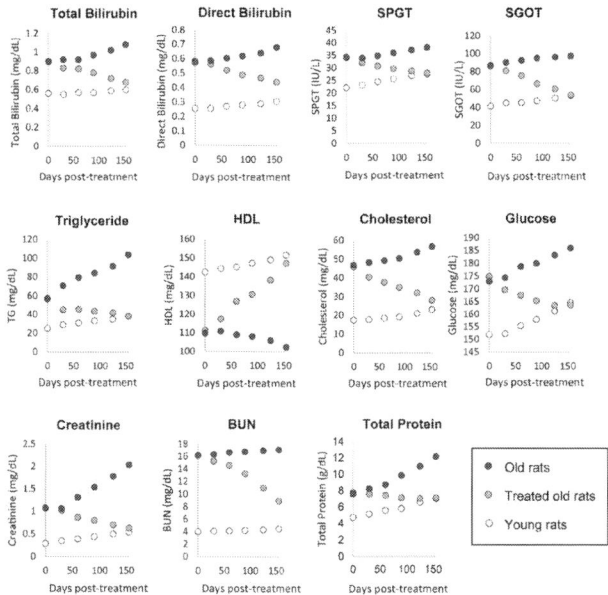

Figure 24: Top Row: liver function — SPGT is an enzyme that gets released into the blood when either liver or heart are damaged, and SGOT is a liver enzyme that leaks into the blood with damage to liver cells. Middle Row: triglycerides are normal fatty acids attached to glycerin, believed to correlate with heart disease. HDL (high-density lipoproteins), "good cholesterol", are molecules that remove cholesterol from cells, while cholesterol refers to total and is a negative correlate of heart health. Bottom row: kidney function — creatinine should be excreted by your kidney; if not, it builds up in your blood. BUN stands for blood urea nitrogen — like creatinine, should be excreted by your kidney; if not, it can be detected by testing for its presence in blood. Adapted from Steve Horvath et al.[1], CC BY-ND 4.0 https://creativecommons.org/licenses/by-nd/4.0/.

Here's what all of the information in Figure 24 means; the experimental group (old, E5-treated rats — each point is the average of six rats) begins at the same levels as the old control rats, and ends with values almost exactly those of the young population after 155 days.

In the graphs in Figure 24, there's no need to figure out how to "connect the dots"; it's obvious which trajectory each group follows. Here again, it's the same color code: dark gray (old), light gray (treated old), and white (young). And these are the same sort of tests that might be given to any aging human, testing liver function, blood lipids (fats) and glucose, as well as kidney functioning. The levels of all clinical detectors of abnormalities in aging organs showed that in our treated animals, all signs of potential liver, kidney, and heart damage were gone. Why? Because your age, your passage through your life-stages, determines all-cause mortality. Nature is out to get you — nothing personal, she's just taking back what is Hers. However, I'm not stealing — it was Nature's gift.

Remember when you could eat anything without weight gain or worry about its health effects? That was then, but may be again. This is a reiteration of the point I made above: the only determinant of destructive or maladaptive changes in the organs and tissues mentioned is age, more specifically, "biological" age, which seems to mean how many of your allotted life-stages have you passed through. And what is the rate at which you pass through them? Clearly, the environment makes a difference, but "environment" (both internal and external) meets genetics through epigenetics. Those maladaptive changes like the increase of LDL cholesterol or the decrease of HDL cholesterol are not so much the result of diet and a sedentary lifestyle as they are the phenotypes that describe later life-stages — phenotypes that include the diseases of aging.

Object #4 - Time to solve the Barnes maze (latency)

The four panels in Figure 25 may not look very interesting, but each point (same color-code) represents the average time it took the six rats of each group to run a Barnes maze (the table with circular holes cut along its periphery, one with a hidden pouch to escape to). "Latency" refers to the time it takes the rat to find the proper hole. One hole is chosen, given a pouch and then markers are placed along the table and walls of the enclosing room to provide guide-posts. The rat is trained for nine days, then tested for nine days. A rat placed in the middle of the Barnes maze table wants to get out of an exposed position, so they naturally look for the pouch to hide in. For each new test, a different hole is chosen to receive the pouch and different markers are placed on the table and walls.

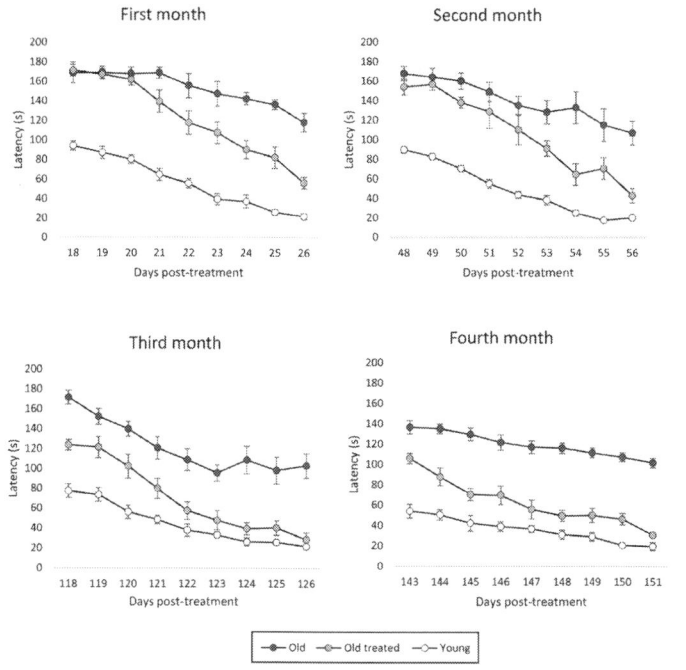

Figure 25: Latency results. Adapted from Steve Horvath et al.[1], CC BY-ND 4.0 https://creativecommons.org/licenses/by-nd/4.0/.

We see at day 18 past the first injection of the first treatment that there's really no difference between the old group and the control group (each group of eight animals was tested every day for nine days). However, by day 21, we see a difference. So, the rejuvenation effect takes longer to manifest for cognitive skills, as expected, because they require more profound changes from the lower biochemical and cellular levels up until changes are made at the organ and organ system levels, in order that the coordination of muscles and neurons linked to the brain activity manifests as improved cognitive performance. Note also that both groups of old rats take about twice the time that young rats do on their initial independent trials, but that by the end of nine days (of the first month), the treated rats had latencies closer to the young group than to the old.

In the second experiment with the Barnes maze (second month), in which the position of the pouch and guideposts were changed, the retrained rats showed a small change at the beginning of the experiments, where their latencies were just a bit lower than the old controls, and the experiment ended with the experimental group even closer to the young controls than at their first trials, though no further E5 was given — meaning that the rejuvenation continued to take place (though by this time inflammatory cytokine levels were starting to rise again). So, the process of rejuvenation seems to take more time to rejuvenate higher functions, but continues to do so even one and a half months (around three human years) following E5 administration.

The second row (months three and four) are those results obtained after the second treatment starting at day 95. You'll notice the experimental rats are considerably more able to solve their maze than the old controls on the first day of testing, though still not so well as the young rats, but as testing continues, on the last day of testing in both graphs (months three and four), treated old rats are nearly indistinguishable from the (now not so) young rats in maze solving (the old rats are considerably heavier than the young rats as well, so that might slow them down).

The point is that memory and problem-solving return to near youthful levels; yet, as the brain is so important, we are planning to devote more time to brain rejuvenation, and have some plans. Of course, while the loss of memory associated with normal aging is worrisome, the loss of your entire life, your family and friends of a lifetime because of dementia (particularly Alzheimer's disease) is the scariest fate for the majority of people, and evidence of brain rejuvenation is welcome news indeed — though there are some other approaches to brain rejuvenation which might be used in conjunction with an E5 treatment.

Pardon me, I should ("want to", because it's a good story) tell you all about E5. However, until we have to reveal them (under the protection of patents), our secrets are our biggest asset. If we gave them away without proper patent protection, we would be left with nothing. Besides that, a real fear of mine is that Big Pharma might want E5 gone, as it might considerably curtail business for them, especially as many of their best-selling — and lifetime — requiring drugs are directed at the diseases and conditions of aging. At some point, we'll have to reveal our secrets, but we also want our twenty years to have exclusive control of where we're headed, and that will be where biology will be headed (images from the movie *Existenz* pop into my head).

Now, some indications of why aging was reversed (or even if it was reversed) were given by the next "object", which discusses multiple items relating to oxidative stress in aging and rejuvenation that required the sacrifice of the animals. An independent laboratory undertook the measurements of organ contents of various molecules that are markers of aging.

Object #5

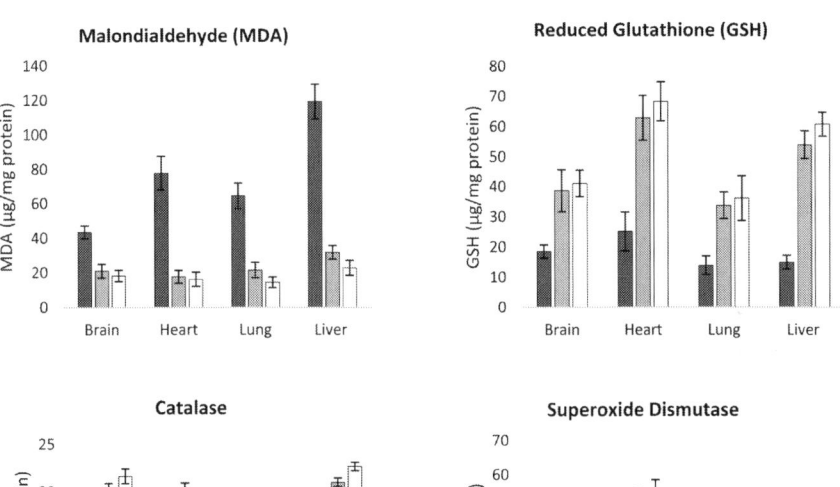

Figure 26: Results of ex vivo analysis of the rats. Adapted from Steve Horvath et al.[1], CC BY-ND 4.0 https://creativecommons.org/licenses/by-nd/4.0/.

As mentioned, the only way to get the readings from Figure 26 was to sacrifice the animals. The individual bar graphs display the organ contents of various redox-connected markers or enzymes involved in the neutralization of oxidative stress, as labeled. Malonaldehyde is a product of fat oxidation and indirectly provides the levels of ROS production by the cells of the organ measured. We see little oxidation in the brain, and more in the liver in old animals as compared to young. But the E5-treated animals were much more like the young controls (around eight months old) than the old controls (22 months old). So, this means that there was less excess ROS production, either because there was rejuvenation of mitochondria (producing fewer ROS per electron transported) or an up-regulation of repair enzymes (as some are known to be down-regulated with aging), or likely both.

The graph to the right of that displays the organ amounts of reduced glutathione, a major determinant of the reducing capacity of a cell's cytoplasm and mitochondria — its "resilience" — as it determines whether or not a cell will survive an oxidative-stress event. The continual decrease in the cellular stores of reduced glutathione determines how far a hydrogen peroxide molecule or high energy nitrogen species might travel across a cell (its "mean free path"), i.e., what its lifetime might be before being quenched by a reducing agent. The greater that mean free path, the more likely it is to do damage.

Glutathione is also said to be a primary determinant of cytosolic and mitochondrial redox potential (whether a molecule will receive hydride atoms or donate them), and that depends on the ratio of the oxidized to the reduced form. Normally, reduced glutathione is written as GSH — "G" is for "glutathione" and "SH" represents the thiol group attached (simply a sulfur atom and a hydrogen atom, GS-H, a group that can lose its hydrogen to become GS- with that empty bond on its right end). Glutathione's reducing action can be described by the reactions 1 and 2 below:

$$2GSH + H_2O_2 \rightarrow GSSG + 2H_2O \quad (1)$$

when reacting with hydrogen peroxide

$$2GSH + R_2O_2 \rightarrow GSSG + 2ROH \quad (2)$$

when reacting with a peroxidated organic compound

Glutathione is at a very high concentration in the cell, but not all glutathione is used as described in the above reactions. Glutathione-S-transferases, for example, are enzymes that bond the entire glutathione molecule to proteins with redox lesions; glutathione becomes enzymatically attached to these proteins and thus they're "cured" (or expelled from the cell). However, the idea that reduced glutathione (GSH) always oxidizes to GSSG is wrong; GSH is used to neutralize toxic chemicals by combining with them, and it is GSH that is responsible for keeping our vitamins C and E in an anti-oxidant state.

So, it is a marker of aging in rats and people that cellular levels of glutathione decrease with aging (levels in the hypothalamus in humans correlate negatively with Alzheimer's disease). Only in some cancer tumors do we see an increase in reduced glutathione — and in the cells of our rejuvenated animals.

So, clearly, though I won't show the data, after about 30 days (our first experiment) the levels of GSH came to about 70% of the values in young rats, but in the graphs in Figure 26 we see that in all organs tested at the end of 5 months and two treatments, levels of reduced glutathione in the organs of the E5-treated old rats were nearly indistinguishable from the young controls, with overlap between the groups in all cases, while no treated rat was anywhere near the values of the old controls. So, instead of gradually losing GSH — a mark of aging in many organisms — the amount of GSH increased per organ, which should give them a greater "organ reserve".

Here, I really felt I was seeing aging reversed, but there was a big surprise yet to come: Steve Horvath's results. Though it seemed to take forever, Steve built his rat clock (Sprague Dawley rats) with our assistance and the additional assistance of Rudy Goya, from Argentina (Rodolfo is his actual first name). So, now, it was time for Steve's part of the bargain — and we didn't seem high on his priority list (I can't blame him), so we waited and waited. I was trying to answer the question of why our E5 would show such dramatic results and Steve's assay would show nothing (just in case), because I was more convinced of our results than the meaning of Steve's assay, but as we'll see, that assay cleared everything up in the best possible way. However, before we get to that, let's see what further information is given by the current "object".

The bottom two bar graphs from Figure 26 are enzymes that have the responsibility to deal with hydrogen peroxide, H_2O_2. Superoxide dismutase produces H_2O_2 from the more energetic and toxic superoxide radical anion. It has mitochondrial and cytosolic versions, and is required for life. Catalase turns hydrogen peroxide into water and oxygen. It is found exclusively on peroxisomes, which are involved in fatty acid catabolism but don't produce ATP (though they do produce NADH). Interestingly, when catalase was moved to mitochondria (genetically, in mice) there was a significant increase in lifespan. So, if instead of hydrogen peroxide you get water and oxygen, and that lengthens lifespan, it would seem hydrogen peroxide (in excess?) contributes to aging and death.

One reason that the levels of antioxidants increased might be shown by the graphs in Figure 27.

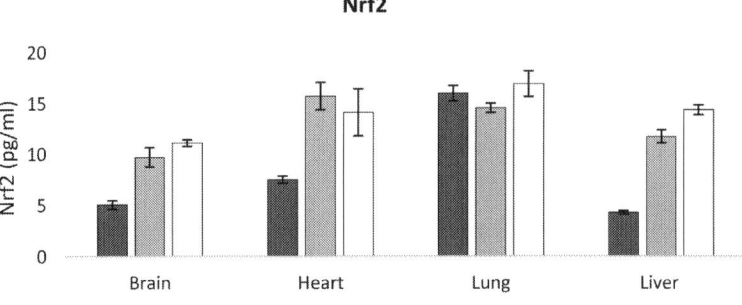

Figure 27: Nrf2 levels obtained in the experiment. Adapted from Steve Horvath et al.[1], CC BY-ND 4.0 https://creativecommons.org/licenses/by-nd/4.0/.

Nrf2 is a transcription factor that is normally held in the cytoplasm and is continuously degraded by the cell; the molecule has a half-life of twenty minutes. But when the cytosol becomes oxidizing, when there's a burst of ROS (reactive oxygen and nitrogen species), the transcription factor is freed from being degraded, builds up its levels and enters the nucleus. There it binds to specific DNA sequences called "antioxidant response elements" (AREs), which are present in the promotor regions of genes and affect their transcription. Many of Nrf2's genes (those it controls) are concerned with repairing oxidative damage and eliminating toxic products of oxidation, such as the genes that code for glutathione-S-transferases. Some of Nrf2's genes are concerned with the production of glutathione by up-regulating glutamate-cysteine ligase, the rate-limiting enzyme in its production.

The thioredoxin family of antioxidant proteins, including peroxiredoxins, are controlled by AREs. So, while in the lungs there was no appreciable difference between treated old rats and untreated, there was also no significant difference from young rats. Otherwise, in the brain, heart and liver, the levels of Nrf2 were at or near youthful levels and clearly different from untreated old animals.

The important thing to note is that although the levels of intra-cellular ROS are increasing with age, the quantity of Nrf2 in organs considerably decreases with age. One possible reason for that is that when ROS levels become very high, the "survival" transcription factor NF-kB takes over. When it does, the cell starts producing inflammatory cytokines, particularly

Il-6 (NF-kB binds to the Il-6 promotor) and anti-apoptotic factors (which prevent it from killing itself — as "cellular etiquette" demands of damaged cells). Another property of NF-kB is that it inhibits Nrf2, but the response is mutual; Nrf2 inhibits NF-kB!

At one point I came up with an explanation of what happened, that went something like this: the cell (cytosol and nucleus) loses reducing potential as it is used up by increasing amounts of ROS and other users of reducing power (synthesis), and it is not replenished quickly enough so that there is a decreasing supply of NADPH and GSH, all caused by the decrease in NAD^+, caused, in turn, by increased DNA repair (as NAD^+ is a substrate for PARP, which is used to mark DNA damage by the laying down of long chains of poly-ADP), and by NAD^+'s usage by sirtuins for deacetylation reactions (remember that in "old" cells, there is a lack of epigenetic maintenance as well as genetic dysregulation). Ultimately, the cell hits a "rough spot" when ROS production or exposure (H_2O_2 penetrates cell membranes) overcomes the resources (NADPH, GSH, NAD^+ and ATP — all components of "resilience") that repair the damage caused, and the cell dies.

I no longer support this model. We must now take into account that the mortality of the organism depends on the life-stage it is in (a proportion of the total lifespan — measured as "mean" (average) or maximal lifespan, as they are proportional), and the fact revealed by the Stanford group (Conboys, Rando, Wagers) that the age of a cell (or more accurately, the age-phenotype — how it looks and acts) is determined by its intercellular environment, by pro- and anti-aging factors in the blood plasma, and not by how long it's lived in the body.[59,61]

The fact that old bodies contain old-appearing cells is because the body controls the cellular age-phenotype, and not because the cells have aged. Cells from centenarians that have been treated with Yamanaka factors (by Laure Lepasset)[68] and brought back to the stemness of embryonic stem cells were able to form any cell type — their telomeres were lengthened, their mitochondria became efficient again as in youth, their transcription profiles went back to youthful patterns — and they could ultimately be used for chimera formation (where they might be mixed with the early embryo of a different species), and could perhaps, if allowed, live another full life.

So, it now very much appears to me that the question as to how our rejuvenation works was answered in the most unexpected and enlightening way.

Object #6

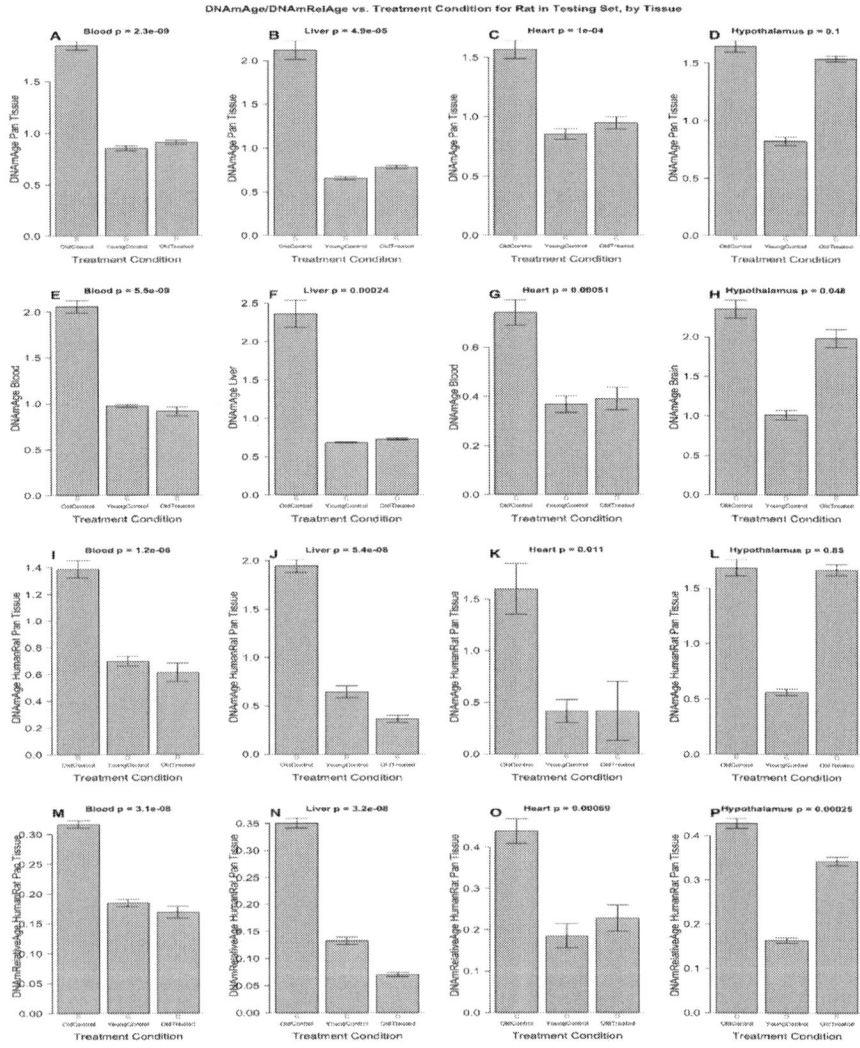

Figure 28: Epigenetic clock analysis of plasma fraction treatment. Each row represents a clock type (six in total), and each column represents a tissue/organ; from the left, blood, liver, heart and hypothalamus. The first row is the rat pan-tissue clock. The second row shows: E) rat blood clock applied to blood, F) rat liver clock applied to liver, G) rat blood clock applied to heart, and H) rat brain clock applied to hypothalamus. The third row represents chronological age as deter-

mined by a human-rat clock (!). The fourth row (graphs M-N-O-P) shows human-rat clock measurements of relative age defined as age/maximum species lifespan. In each graph, the first bar (from left to right) is the old control, the second bar is the young control and the third bar is the old treated rats. Image by Steve Horvath et al.[1], CC BY-ND 4.0 https://creativecommons.org/licenses/by-nd/4.0/.

On March 16th, 2020 I received an e-mail from Steve Horvath with the subject bar bearing the title "fantastic results rat clock" and he said our treatment worked; in fact we had better than halved the epigenetic age (DNA methylation age) of our old treated rats by several of Steve's clocks (including a pan-species clock which we'll discuss). In figure 28, each row represents a clock type, as explained in the legend.

The obvious unconforming result is the smaller amount of rejuvenation in the hypothalamus as compared to other organs. However, maze learning showed a vast improvement in performance over old controls and equaling young controls by the end of a testing cycle. So, the meaning of this apparently lessened rejuvenation of the hypothalamus is unclear, but we will clearly look further into this. I've already got ideas.

At this point, even knowing as little as we do, we can re-program cells in a "natural" and hopefully all-inclusive way that will solve the aging problem. E5 will be the first step. This is no dream — Steve's assays confirmed all of our other assays in which age-dependent (and some traits previously unknown to be so) levels of biochemical, physiological and cognition returned to youthful levels. Thus, the rise in good cholesterol (HDL) vs. bad cholesterol (LDL) is not a function of lifestyle but of life-stage.

Stroustrup et al.'s observation that all-cause mortality is basically a function of life-stage (as a defined proportion of total lifespan) is coupled together with David Neill's theory of life as a progression of life-stages, and Steve Horvath's Epigenetic Theory of Aging, as I will explain — but the ultimate reason for how and why E5 works is simple enough; plasma is the body's way of controlling the age-phenotype of its cells, and E5 returns cells to an earlier epigenetic state. The now rejuvenated cells begin to repopulate the depleted tissues (our treatment appears to remove senescent cells as well), the reinforced tissue now contributes to better functioning organs, stronger hearts, more elastic arteries, less permeable intestines and kidneys. That, of course, strengthens us. So far, we have repeated this experiment successfully many times.

9

The new science of rejuvenation

In general, according to the "wear and tear'ists", "rejuvenation" should be impossible. Damages can be vital — damages would have been repaired if they could have been repaired. The aged cell is a damaged and irreparable cell. Work by Briggs and King and J.B. Gurdon in the 1980s with somatic cell nuclear transfer (SCNT) shows that even the adult (in some cases) stem somatic skin cells had nuclei which when implanted resulted in complete animals that went all the way to adults. The more recent cloning activities were not limited to frogs and toads (Gurdon used the squat and sluggish South African clawed toad, *Xenopus laevis*, as his experimental subjects).

Dolly the sheep is another example that the nuclei of the cells of mature animals are capable of forming fully functional animals. Also, in an experiment that started in 2000, calves that resulted from the SCNT of the nuclei of cow fibroblasts grown to replicative senescence in vitro (in cell culture) became fully grown cows, all perfect so far.[69] In that experiment, the cells of a cow were cultured in vitro until they could no longer multiply (replicative senescence). This senescence was once deemed too degraded to support life, much less normal development. Nuclei were extracted from some of those now senescent cells and placed into cow ova from which the nuclei had been removed, and then those ova were implanted in cows and allowed to come to term.

Figure 29: South African clawed toad, *Xenopus laevis*. Brian Gratwicke, CC BY 2.0 https://creativecommons.org/licenses/by/2.0, via Wikimedia Commons

Somatic cell nuclear transfer showed that even the nucleus of cells aged both in vivo and in vitro could give rise to normal offspring. This unambiguously demonstrated that the purportedly "worn and torn" old nucleus still had what it took (gene-wise) to form a completely normal animal (with many kinds of animals now produced in this manner; it's become part of commercial animal breeding). So, if it was not the nucleus that was aged, then was it the cytoplasm? Or did the cytoplasm of the enucleated egg cell have a powerful restorative effect on the aged and apparently senescent nucleus?

The answer came from an unexpected direction — a convergence of two very different methodologies revealed it. If a cell is treated with certain "pioneer transcription factors", it will change its state of differentiation. Though only a small proportion of cells were "prone" to this change, the change turned ordinary somatic cells into what appeared to be embryonic stem cells, capable of forming any cell type (except the extraembryonic membranes of the early embryo — which later becomes part of the placenta). And with the proper direction by tissue-specific transcription factors inserted as plasmids, those *induced pluripotent stem cells* (iPSCs), as they were called after the transformation, could be made into any cell type desired (though they still have not been found capable of fully forming an animal by themselves as SCNT cells can).

Furthermore, when cells were set back to iPSCs, they were also set back to age zero, and all "cellular aging" characteristics were reversed. At the same time, if cells are transdifferentiated by direct means that by-pass cells going through the iPSC stage, the aging "clock" is not reset, only the patterns of cell differentiation and the epigenetic mechanisms that control them are.

So, is there some distance along the path from differentiated somatic cells to de-differentiated iPSCs where only age changes? And it turns out that with controlled administration of the Yamanaka factors, these "pioneer" transcription factors allow cells to change their states and set back their epigenetic age. The original Yamanaka factors, known as OSKM (Oct 4, Sox 2, Klf4, c-Myc) produce rejuvenation when expressed in limited, sequential time periods, but in several experiments — including in vivo experiments with mice — there was the unfortunate side effect of the formation of cancers with teeth (and other organs or parts thereof), called teratomas. Recently, David Sinclair, using an in vitro system with only OSK (as M was considered responsible for carcinogenesis), regrew an optic nerve in vitro from an old eye.[70]

E5 should have all these properties, but should diminish the incidence of cancer as cancer is a disease of old age, and the reprogrammed cells should respond the way they would have when they were young. Perhaps, except that now stem cells will be more engaged in self-renewal to increase their numbers — and their numbers determine the numbers of their differentiated progeny, to solve the cytopenia problems (loss of tissue cells) of old animals. It would seem that E5 affects many sorts of pioneer transcription factors in tissues: it changes the cellular age phenotype in place, by simple injection, with no need to engineer ourselves or our children. However, I do acknowledge that some scientists, like David Sinclair, are beginning to see the truth about aging. I hope I helped — I am sure that E5 will be the beginning of a new and correct biology that will put fact before theory.

So, we are not alone in realizing that cells can be reprogrammed and rejuvenated. Ocampo et al. carried out the most convincing demonstration that partial reprogramming using the transient and sequential expression of the OSKM genes in living mice can significantly increase their lifespan, plus bringing to youthful levels all the other biological age-markers examined.[71] All four OSKM factors were required, but the point was proven: even in vivo, in situ (in a living animal, with everything "in place") significant systemic rejuvenation occurred by invocation of these pioneer transcription factors — which first bind to DNA, allowing other transcription factors and co-activators to bind later and thus allowing chromatin remodeling and differentiation, an event which lowers the reducing potential of the cell, or vice versa.

Blood factors

In their summary paper of new efforts in anti-aging research, Mahmoudi, Xu and Brunet discuss four potential methods of treating aging,[72] with our research indirectly included (it wasn't specifically mentioned, but I told Anne Brunet about our work at the 2019 BAAM, Bay Area Aging Meeting, at Stanford). The first is blood factors, where they talk about the usual suspects: pro-aging factors in old blood plasma. However, while the Conboys showed that diluting old blood plasma has an apparent antiaging effect judging from the appearance of tissues before and after treatment,[52] E5, which is 100% blood composed, has a much greater and more permanent effect without significant dilution (2% every other day for four days), showing that the youth-promoting factors in young blood have a more permanent effect and seem to be dominant over age-promoting factors. Following our second E5 treatment of rats, age was reset followed by aging at normal rates. I will discuss why after I discuss Brunet group's other three possibilities.

The authors follow up with a very nice table showing the effects of heterochronic parabiosis (HPE in case you forgot — my term) and the effects of other blood-derived products (oxytocin, TIMP2, gdf11) on rejuvenation. It is of note that simple blood injection improves cognition.

Metabolic interventions

The next area the group talks about is metabolic interventions, in terms of diet or chemicals like the mTOR-C1 complex inhibitor rapamycin. To begin with, this is not an approach that leads to immortality; at best, in mammals at least, it leads to a fractional increase in lifespan. Caloric restriction is known to work to extend lifespan in every group, vertebrate and invertebrate, except perhaps primates, where tests of caloric restriction did not show increased lifespan, though they did show increased healthspan.

That caloric restriction works was shown in the 1930s by Clive MacKay (who also used heterochronic parabiosis to study aging).[73] However, the long-term studies by The National Institute on Aging (USA), using rhesus monkeys, did not show lifespan extension — while other studies did. The real point here is something more obvious: if the effect existed, why did it require statistics to know if the experiment succeeded or failed? If statistics were required, then the effect was insubstantial. To me, the critical part of caloric restriction and the related methionine restriction is that they are aimed at extending cellular life by reducing metabolic activity.

I believe the speed of the "clock" is set by the amount of damage produced; in adulthood, there seems to be a sequential down-regulation of required proteins throughout the lifespan, occurring in different organs at different times. The thymus starts involuting (getting smaller and turning to fat) early, in childhood, in all people, while in women the ovaries age and become dysfunctional past mid-life. Other organs age more slowly. Even cells of the same type, that age by the same mechanisms, age at different rates in different organs, leading to my belief (for what it's worth) that cell state (including age-phenotype) is determined at the organic level (the level of the organs), as well as the systemic level (which may act through the organs).

My supposition (actually Stroustrup's conclusion) is that damage-inducing conditions cause accelerated aging and are therefore responsible for death by all causes; then, my model, based on Neill's model, but with life's timer being the result from the daily loss of "resilience" — the measure of which are the concentrations of GSH (reduced glutathione), thioredoxins, NADPH and NAD^+ — indicates that while it cannot disturb the fixed circadian cycle (as there are many external zeitgebers ["time givers"]), it does change the rate of damage accumulation and hence the lengths of all life-stages.

As stated before, even when this caloric restriction diet started in early adulthood, it could delay development (adult "development") and could result in an extended lifespan, including an expanded middle-age and expanded old age. However, even if it would benefit the individual and society by its health outcomes, it is not a road to immortality or even to rejuvenation, as it only delays the inevitable (which is all for the good), by "slowing the clock". The authors Mahmoudi et al. state that the "mode of action" of metabolic interventions includes "blood factors" (certainly including E5 then),[70] and certainly there are blood factors that change metabolic states of cells (hormones, for example), but it is not my belief that E5 works in this manner, as I will discuss.

A note about rapamycin

Rapamycin is an interesting drug. Back in the day when explorers visited an unexplored territory, they always took soil samples to send back to laboratories to test for microbes that produced novel antibiotics. Well, they hit pay dirt (excuse me) when they tested the soil of the Pacific island famous for the huge stone gods that are set around it to protect the inhabitants from invasion from the sea. Easter Island, off the west coast of South America, is called Rapa Nui by its Polynesian inhabitants. The secretion of a product by the soil bacterium found there (*Streptomyces hygroscopicus*) was immunosuppressive and had anti-cancer effects. Further work elucidated the target of rapamycin as the protein mTOR (originally "mammalian target of rapamycin").

Here's the thing, though; rapamycin is an inhibitor of an enzyme, a kinase called mTOR (remember, "kinases" are enzymes that attach phosphate groups to other molecules including proteins, activating or inhibiting or even changing the function of the proteins they phosphorylate). The very name mTOR means "mechanistic Target Of Rapamycin", so the drug rapamycin regulates the regulator of cell metabolism.

The mTOR Complex 1 (mTORC1) is our concern here, as it has inputs from many aspects of cell function, including energy levels, synthetic activities and growth. As in the yin-yang symbol, besides the light-side — a growth-oriented, energy-producing metabolism controlled by the mTORC1 — there is a dark side with opposite functions, in this case the FOXO (forkhead family of transcription factors) with the opposite mission, to slow down growth and energy production and to initiate repair and maintenance.

When FOXO dominates, other subsidiary repair and maintenance transcription factors like Hsp1 and Nrf2 are also transcribed, each with their own, as well as shared, repair and maintenance activities. These repair and maintenance activities have been shown to increase lifespan, as have many mutations shown to produce a lesser rate of oxidation damage (and often at the cost of reduced growth and activity) or damage by protein misfolding (a "hallmark" of aging — due in part, I believe, to the reducing condition of the endoplasmic reticulum when it should be oxidizing).

I actually tried rapamycin at what should have been an effective dose, but didn't find any significant effect until the rapamycin therapy interacted with my intestinal bowel disease (IBD); the same process that works against growth in most somatic cells works against growth in the cells needed to repair the intestinal damage done by IBD, and so I would absolutely not recommend it for people with that or any condition necessitating tissue growth (which of course is why it is an anti-cancer drug) and repair. However, rapamycin treatment is at best only another means of fractionally extending mammalian lifespan, by extending old age — but with risks (including a weakened immune system) — and that's better than nothing, I suppose (if you are not too incapacitated by old age). However, it's merely a side trail leading back to death after a tortuous and dangerous journey.

Removal of senescent cells

The third way reviewed in Mahmoudi et al.'s article is the removal of senescent cells. Since their discovery, Judith Campisi has studied senescent cells extensively, and showed that rather than the benign and nearly dead cells, as they have been taken out of the cell cycle (they don't reproduce) and perform no known useful work (though they do appear to be involved in wound healing), they are in fact metabolically active zombie cells that actually harm the body by producing inflammatory cytokines and extracellular proteinases, collagenases and gelatinases that break apart the intercellular matrix that holds our cells together, and makes it easy for migrating cancer cells to reach the blood and lymphatic vessels through which they spread.

Darren Baker sought, by means of genetic engineering, a treatment that would preferentially kill senescent cells. He chose as his target the elaboration of the protein $p16^{INK4a}$ (named for its ability to inhibit NF-kB), a marker of cellular senescence. $p16^{INK4a}$ drives cells into senescence and so is a very important tumor suppressor gene. Baker designed a transgene, INK-ATTAC, that allowed for the elimination of cells expressing $p16^{INK4a}$ by the addition of an exogenous substance (an antibiotic).[74]

In the experiment, one mouse periodically had its senescent cells eliminated ($p16^{INK4a}$ expressing cells), and another mouse did not. The differences in appearance were startling (the mouse which had senescent cells eliminated looking much healthier and younger). However, there was no significant difference in longevity (this was a short-lived, progeroid strain). Baker performed the same experiment with a wild-type mice (with normal lifespans) and found the same effects and a fractional increase in lifespan.[72]

Clearly the removal of senescent or at least p16^{INK4a} expressing cells increases lifespan and decreases the incidence of morbidities associated with aging. However, it is certainly yet another side road that inevitably leads us back to aging and death, but looking and feeling better. Still, a worthy goal; I've taught a Biology of Aging course for years and I am constantly surprised that many people have no desire for extended lives — actually, not so surprised; without Heaven and constant God-constructed entertainment, life can be dull. It was an ancient Christian myth that a Jew who, when Jesus was carrying his cross on the "Via Dolorosa" and stopped to rest, demanded of him "Why do you tarrying here?" was for that condemned to live until Jesus returned. This was considered a harsh punishment. And what I do find is the more common response to the question about the desire of living extended lives is living to a "ripe old age" (no cannibalism here) in good health and dying in one's sleep. Not entirely a plan without merits if you can't do better (or think that one life is enough and more than enough — as do many on this planet).

In any case, we have some evidence (though not great evidence) that E5 treatment eliminates senescent cells, as we had a distinct lack of senescence-associated beta-galactosidase staining, a marker for senescent cells, but not a definitive marker — it usually takes another marker, like p16^{INK4a} antibodies (conjugated to a fluorescent compound to make it visible), to define senescent cells. Even so, senolytics — treatments that eliminate senescent cells — are not a pathway to immortality, though they do have some characteristics of rejuvenation and it appears as though they would improve life, especially old age, for those who could eliminate senescent cells. One factor I forgot to mention about senescent cells' toxic secretions is that they secrete an unknown substance that appears to convert other, neighboring cells to senescence.

Epigenetic reprogramming

The final category of Mahmoudi et al.'s paper is epigenetic reprogramming. While this still has problems, the work of Ocampo clearly demonstrates the potential.[69] The use of OSKM leads to teratomas (not what you would want), but it is clear from these demonstrations that aging is an epigenetic phenomenon, or at least controlled through epigenetic means. And it is here where I would place E5, rather than among the blood products mentioned. The reason for that is simple. Steve Horvath's clocks showed that not only we had rejuvenated many tissues and organs (as when looked at closely [very closely], not only was the phenotype of all cells and organs examined youthful), but also the epigenomes (at least those parts that mark or determine age), which were modified to be what would be expected of animals whose biological age is less than half their chronological age. So, the DNA methylation profiles confirmed that the cells had been rejuvenated at the very deepest level — their epigenomes. As this is the case, and as re-treatment appears to work better than initial treatments — leaving aging to continue at normal levels — it should be possible to keep an animal at a young adult life stage indefinitely, so long as E5 is continually provided.

What is a life?

In 1944 (my birth year), Erwin Schrödinger wrote the short book *What is life?*,[75] a classic book that influenced many physicists to go into the biological sciences. Schrödinger, a prominent physicist, concluded that life must be organized by an "aperiodic" crystal; an "almost" crystal, with nearly identical subunits — molecular constituents — arranged in order to be held together by covalent bonds. This was written before the discovery of the chemical nature of DNA by Crick, Watson, Pauling, and belatedly Rosalind Franklin, the scientist who the "boys" were so afraid of, and whose X-ray diffraction studies of DNA resulted in the discovery of the double helix, base-pairing complementarity — it's always a cytosine paired with a guanine, and an adenine is always paired with a thymine residue on the strand opposite it. Specifically, Franklin independently determined the double-helix as the result of her X-ray diffraction pattern.

However, if you noticed, the name of this section is "What is **a** life?" The point of this is to acknowledge as a first principle that life is not a process of development that ends leaving a living sexually mature organism that continues to live until it hits some bump on life's highway it is unable to overcome (usually being eaten by predators, for herbivores, or starving, for carnivores — cold is the biggest natural killer of mice). In the case of the workers of honeybee hives, their lifespans depend on when they were born (in places with seasons), which determines when they leave the nest. The life of a bee (like that of a T4 virus) plays out with little variation, as though from one end to another a series of stereotypical roles are played, first as a nursemaid (all worker bees are maidens), then through a variety of life-stages that determine their function within the hive — repairing damage, building, or, in the case of bees in the winter, furiously beating their wings to heat the air within the hive, keeping it comfortable for the queen and her entourage. But finally, at the end of their lives, bees enter a state called "forager", with a much higher intrinsic mortality, and foragers die within a few days.

The point is that organisms are given a life — not the process of life that might end if it meets more than it can handle, or a life plan determined by chance; they're a four-dimensional object, limited in time as well as in space with many mechanisms designed to ensure that lifespan does not pass the species maximum lifespan, as lifespan is a species characteristic, like size or coloration, that depends on an organism's ecological niche and habitat as well as many other physical and biological "facts of life". Once this is realized, the "mystery" of aging disappears; why do "things fall apart"? They were made that way — built-in obsolescence is a trick that human manufacturers got from Nature, and oddly enough, for the same reasons, to promote demand and seed innovation, as people are always looking for a "new and improved" product.

The most important point I have to make, and one that changes everything (but there's more to come), is that "cellular aging" is a myth; the cell's age is determined by its cellular environment, and not by its history. The aging of the organism is not determined by the aging of its cells — its cells' age-phenotype is determined by the biological age of the organism.

We have learned from Albert Einstein that time is not a constant that flows everywhere at the same rate, but in organisms, the flow of time, as evidenced by the passage through their lives, is determined by factors other than mass and density. What we demonstrated is that time — being the sequential ordering of events in a local environment — has a totally different meaning in biological systems, where time, in terms of passage through their lives, depends on an organism's ecological niche.

So, biological systems are very unlike physical systems, as "biological time" can be slowed, stopped or even reversed without any contradiction to the laws of physics. The second law of thermodynamics, "entropy", "doesn't apply" to living systems, as they gain negentropy[*] with time; biological systems are open systems — they receive and transmit both energy and mass.

---

[*] In information theory and statistics, negentropy is used as a measure of distance to normality. The concept and phrase "negative entropy" was introduced by Erwin Schrödinger in his 1944 popular-science book What is Life? Later, the phrase was shortened negentropy.

To imagine that such a system would have to obey the laws of entropy as if it were a closed system would be equivalent to saying air-conditioning won't work because warmth flows to colder areas. However, just as in living things, the application of energy can reverse the effects of entropy.

So, organismic aging according to species lifespans seems to be very like the model of David Neill, a fixed succession of life-stages timed by a redox clock based on the circadian cycle and the corresponding sleep-wake cycle. Blood plasma (in mammals at least) determines the age-phenotype at the cellular level, which subsequently has been shown to be reversible, and once reversed at the cellular level, those changes will ascend the biological hierarchy, from cells, to tissues, organs, organ systems, and ultimately the whole body to produce a rejuvenated organism.

My ultimate conclusion is that lifespan is a controlled succession of life-stages, lasting throughout adult life and ending in death, but that this is not a physical law but a biological one, reinforced by the evolutionary forces (including group selection), in response to the role in the ecosystem of which the organism is a part. The basic thesis is that lifespan is a hereditary species characteristic, available to small modifications, or very much larger ones in much the same manner as weight or skin patterning, and controlled by a genetic regulatory network of ancient provenance with homologous mechanisms throughout most animal phyla in the world.

One unique study made with the Madagascan "mouse lemur", *Microcebus murinus*, a very small lemur with a great sensitivity to seasonal changes in the photoperiod, showed that when exposed to altered day-night cycles (in an enclosed environment), the lemurs responded to cycles 2 ½ times the frequency by aging at a faster rate, establishing at least in these animals the correspondence between aging and circadian rhythms.[76] This also explains the results of the Stroustrup and Fontana experiments on large numbers of *C. elegans* with unprecedented accuracy.[39] The lemurs responded to a speeded-up day-night cycle by speeding up the transition to later life stages.

Now what?

Of course, there are many, many more questions than answers, and you might well ask how we can use this invention and discovery when we don't have the slightest idea of how it works (though I know approximately what it IS). The answer is that electrical technology, motors and telegraphs, and even lighting (in arc lamps), were used before we knew what electrons were. E5 therapy could be applied at any age, but will be reserved for the aged until supplies exceed demand by that group.

There are fundamental questions that need answers. For example, is there simply one kind of E5, as it is very clear from the recent experiments of the Conboys and Kiprov I mentioned that there are pro-aging factors in the blood of older animals?[59] (Though now the evidence supports Kiprov's contention that young serum albumin is a pro-youth agent.[15]) So, does that combination of pro- and anti-aging factors result in definite age-signatures of the varying tissues? That is, it is clear that the composition of the blood plasma determines the age-phenotype of cells in tissues, but we have no idea of how.

As we know that the DNAm age changes with adult chronological age (which is the basis of all DNAm clocks), we already know that old plasma will age cells, and though it has yet to be done (we will do it), old plasma should change (increase) the DNAm age of the cells of a young organism treated with it. That alone could be the most frightening torture in history — to take a young person and make them old, which is already the theme of several sci-fi dramas I've seen. However, the nice part is that there appears to be no reason that a person couldn't become young again with an E5 treatment, as it has already been shown that E5 works in the presence of old plasma, so that nothing as drastic as plasma exchange needs to occur (though it might be useful to speed up rejuvenation) — maybe some hybrid technique.

So, if E5 works as well on humans as on rats, we should be able, in the course of several years, to return all our important cells, tissues, organs and organism to youth. If factors are lacking, we will find them. I believe our discovery will help to elucidate the development in both pre- and post-maturity. The way I see the future, at least in the beginning, is to have those who wish, using it — as many may object for religious reasons (at least one Pope inveighed against life-extension by artificial means, and this is surely an artificial means).

To me though, to quote the great Galileo, "The Bible shows the way to go to heaven, not the way the heavens go." And the real book of God's creation is, as Galileo told us, "written in the language of mathematics". But mathematics is a tool for clear thinking about concrete problems, while biology is an assemblage of a complexity beyond our ability to grasp but in the loosest terms. We shall soon allow the world a glimpse of a complexity hidden from our eyes, an unknown world with the potential to give us more than all the gold, silver and diamonds combined: life.

What will be the repercussions of E5 to human society? That's a question above my pay grade. The process of rejuvenation seems to take months, if not years. There is a one-time injection or infusion of E5 — which in the past was administered over a week, but for technical reasons related to injecting rats with their fragile tail veins, and there is no reason the entire amount couldn't be administered during one or two (to be safe) office visits. The effects on physical strength, acuity and inflammation should kick in almost immediately, with decreasing levels of pro-aging factors and increasing levels of anti-aging factors derived from young adult plasma. As we find that the effect of E5 is to change the cellular age-phenotype, the pro-aging factors produced by cells with a later age-phenotype will no longer be produced, and eventually rejuvenation will be complete — at least the rejuvenation of major vital organs, including brain, heart, lungs, liver and kidneys, in terms of biochemistry and performance. We have no idea about cancer, but as age inversely relates to cancer rates, E5 should prevent cancer from occurring, but how it will act on cancer cells (will it "youthenize" them?) remains to be seen.

For certain, we already have target diseases of aging, low-hanging fruits with no competition, but ultimately aging is itself our target. I believe in the ultimate vision seen in the Bible (which has neither mention of, nor belief in an "immortal soul"): humankind immortal in Heaven, though in ways the ancients couldn't imagine.

References

1. **Reversing age: dual species measurement of epigenetic age with a single clock**
Steve Horvath, Kavita Singh, (…) Harold L. Katcher
bioRxiv 2020.05.07.082917.

2. **The serial cultivation of human diploid cell strains**
L. Hayflick and P.S. Moorhead
1961 Experimental Cell Research 25(3): 585-621.

3. **Ending Aging: The Rejuvenation Breakthroughs That Could Reverse Human Aging in Our Lifetime**
Aubrey de Grey with Michael Rae
2008 Griffin.

4. *The Holy Bible, New International Version*®, NIV® Copyright © 1973, 1978, 1984, 2011 by *Biblica, Inc.*® Used by permission. All rights reserved worldwide.

5. **Death and Grief in the Greek Culture**
Kyriaki Mystakidou, Eleni Tsilika, et al.
2005 OMEGA - Journal of Death and Dying 50(1): 23–34.

6. **Entropy Explains Aging, Genetic Determinism Explains Longevity, and Undefined Terminology Explains Misunderstanding Both**
Leonard Hayflick
2007 PLOS Genetics 3(12): e220.

7. **The RNA World: molecular cooperation at the origins of life**
Paul G Higgs and Niles Lehman
2015 Nature Reviews Genetics 16: 7–17.

8. **Complex archaea that bridge the gap between prokaryotes and eukaryotes**
Anja Spang, Jimmy H Saw, et al.
2015 Nature 521:173–179.

9. **Isolation of an archaeon at the prokaryote–eukaryote interface**
Hiroyuki Imachi, Masaru K. Nobu, et al.
2020 Nature 577: 519–525.

10. **The Concept of Evolution to 1872**
Phillip Sloan
The Stanford Encyclopedia of Philosophy (Fall 2018 Edition), Edward N. Zalta (ed.), Available at: https://plato.stanford.edu/archives/fall2018/entries/evolution-to-1872/

11. **The Soul of Culture Vol. 1**
William Anderson Gittens
2019 Devgro Media Arts Services

12. **Descartes' Myth**
Gilbert Ryle
1949 In The Concept of Mind. London: Hutchinson, 11-24.

13. **How Islam changed medicine**
Azeem Majeed
2005 BMJ 331(7531): 1486–1487.

14. **Rejuvenation of three germ layers tissues by exchanging old blood plasma with saline-albumin**
Melod Mehdipour, Colin Skinner, et al.
2020 Aging (Albany NY), 12: 8790-8819.

15. **Young and Undamaged rMSA Improves the Longevity of Mice**
Jiaze Tang, Anji Ju, et al.
2021 bioRxiv 2021.02.21.432135.

16. **The origin and early evolution of eukaryotes in the light of phylogenomics**
Eugene V Koonin
2010 Genome Biology 11: 209.

17. **Leaf Senescence in Wheat: A Drought Tolerance Measure**
Hafsi Miloud and Guendouz Ali
Plant Science - Structure, Anatomy and Physiology in Plants Cultured in Vivo and in Vitro
Ana Gonzalez, María Rodriguez and Nihal Gören Sağlam
IntechOpen. Available at: https://www.intechopen.com/chapters/71539

18. **The Naked Sun**
Isaac Asimov
1972 New York: Fawcett Crest

19. **Ciliate Genome Sequence Reveals Unique Features of a Model Eukaryote**
Richard Robinson
2006 PLoS Biol 4(9): e304.

20. **An amicronucleate mutant of** *Tetrahymena thermophila*
Anthony R Kaney and Virginia J Speare
1983 Experimental Cell Research, 143(2): 461-467.

21. **Siliceous deep-sea sponge** *Monorhaphis chuni*: **A potential paleoclimate archive in ancient animals**
Klaus Peter Jochum, Xiaohong Wang, et al.
2012 Chemical Geology 300–301: 143-151.

22. **Age induced mutations in Paramecium**
T M Sonneborn and M Schneller
1960 The biology of aging, Waverly Press Baltimore

23. **Age-correlated changes in expression of micronuclear damage and repair in** *Paramecium tetraurelia*
Steven R Rodermel and Joan Smith-Sonneborn
1977 Genetics 87(2): 259-274. PMID: 924139, PMCID: PMC1213739

24. **DNA repair and longevity assurance in Paramecium tetraurelia**
Joan Smith-Sonneborn
1979 Science 203: 1115-1117.

25. **The Descent of Man, and Selection in Relation to Sex**
Charles Darwin
1871, 1st ed. London: John Murray.

26. **The Selfish Gene**
Richard Dawkins
1976 New York: Oxford University Press.

27. **Evolution of lifespan**
David Neill
2014 Journal of Theoretical Biology 358: 232-45.

28. **Life-history connections to rates of aging in terrestrial vertebrates**
Robert E Ricklefs
2010 Proceedings of the National Academy of Sciences of the United States of America 107(22): 10314-9.

29. **Evolutionary theories of aging: confirmation of a fundamental prediction, with implications for the genetic basis and evolution of life span**
Robert E Ricklefs
1998 Am Naturalist, 152(1): 24-44.

30. **An unsolved problem of biology**
Peter B Medawar
1952 HK Lewis and Co.

31. **The Essence of Aging**
Jan Vijg and Brian K Kennedy
2016 Gerontology, 62(4): 381-5.

32 **From rapalogs to anti-aging formula**
Mikhail V Blagosklonny
2017 Oncotarget 8(22): 35492–507.

33. **Epigenetic clocks reveal a rejuvenation event during embryogenesis followed by aging**
Csaba Kerepesi, Bohan Zhang, et al.
2021 Science Advances 7(26): eabg6082.

34. **Beta-carotene and lung cancer in smokers: review of hypotheses and status of research**
Regina Goralczyk
2009 Nutr Cancer 61(6): 767-74.

35. **A proposal in relation to a genetic control of lifespan in mammals**
David Neill
2010 Ageing Research Reviews 9: 437–446.

36. **A *C. elegans* mutant that lives twice as long as wild type**
Cynthia Kenyon, Jean Chang, et al.
1993 Nature 366: 461-464.

37. **Comparison of mitochondrial pro-oxidant generation and anti-oxidant defenses between rat and pigeon: possible basis of variation in longevity and metabolic potential**
Hung-Hai Ku and R S Sohal
1993 Mechanisms of Ageing and Development 72(1): 67-76.

38. **Evolutionary Theories of Aging: Confirmation of a Fundamental Prediction, with Implications for the Genetic Basis and Evolution of Life Span**
Robert E Ricklefs
1998 The American Naturalist 152(1): 24-44.

39. **An Analysis of the Relationship Between Metabolism, Developmental Schedules, and Longevity Using Phylogenetic Independent Contrasts**
João Pedro de Magalhães, Joana Costa and George M Church
2007 The Journals of Gerontology: Series A 62(2): 149-160.

40. **The physiology/life-history nexus**
Robert E Ricklefs and Martin Wikelski
2002 Trends in Ecology & Evolution 17(10): 462-468.

41. **The temporal scaling of *Caenorhabditis elegans* ageing**
Nicholas Stroustrup, Winston E Anthony, et al.
2016 Nature 530: 103–107.

42. **The Hallmarks of Aging**
Carlos López-Otín, Maria A Blasco, et al.
2013 Cell 153: 1194-1217.

43. **The Hallmarks of Cancer**
Douglas Hanahan and Robert A Weinberg
2000 Cell 100(1): 57-70

44. **Increased Wnt Signaling During Aging Alters Muscle Stem Cell Fate and Increases Fibrosis**
Andrew S Brack, Michael J Conboy, et al.
2007 Science 317(5839): 807-810.

45. **Cytoplasmic and Mitochondrial NADPH-Coupled Redox Systems in the Regulation of Aging**
Patrick C Bradshaw
2019 Nutrients 11(3): 504.

46. **Age-related changes in the glutathione redox system**
Mine Erden-İnal, Emine Sunal and Güngör Kanbak
2002 Cell Biochemistry and Function 20: 61-66.

47. **Aging effects on DNA methylation modules in human brain and blood tissue**
Steve Horvath, Yafeng Zhang, et al.
2012 Genome Biology 13: R97.

48. **Hypothalamic programming of systemic ageing involving IKK-β, NF-κB and GnRH**
Guo Zhang, Juxue Li, et al.
2013 Nature 497: 211-216.

49. **Cross-talk between circadian clocks, sleep-wake cycles, and metabolic networks: Dispelling the darkness**
Sandipan Ray and Akhilesh B. Reddy
2016 Bioessays 38: 394–405.

50. **Redox characteristics of the eukaryotic cytosol**
H Reynaldo López-Mirabal and Jakob R Winther
2008 Biochimica et Biophysica Acta (BBA) - Molecular Cell Research 1783(4): 629-640.

51. **Circadian Rhythms and Sleep in *Drosophila melanogaster***
Christine Dubowy and Amita Sehgal
2017 Genetics 205(4): 1373–1397.

52. **Role of Nicotinamide Adenine Dinucleotide and Related Precursors as Therapeutic Targets for Age-Related Degenerative Diseases: Rationale, Biochemistry, Pharmacokinetics, and Outcomes**
Nady Braidy, Jade Berg, et al.
2019 Antioxidants & Redox Signaling 30(2): 251-294.

53. **SIRT2 induces the checkpoint kinase BubR1 to increase lifespan**
Brian J North, Michael A Rosenberg, et al.
2014 The EMBO Journal 33(13): 1438-53.

54. **NAD^+ and sirtuins in aging and disease**
Shin-ichiro Imai and Leonard Guarente
2014 Trends in Cell Biology 24(8): 464-471.

55. **Rejuvenation of aged progenitor cells by exposure to a young systemic environment**
Irina M Conboy, Michael J Conboy, et al.
2005 Nature 433: 760–764.

56. **Heterochronic parabiosis: historical perspective and methodological considerations for studies of aging and longevity**
Michael J Conboy, Irina M Conboy, Thomas A Rando
2013 Aging Cell 12(3): 525-30.

57. **Parabiosis between Old and Young Rats**
Clive M McCay, Frank Pope, et al.
1957 Gerontologia 1: 7–17.

58. **Mortality in Syngeneic Rat Parabionts of Different Chronological Age**
Frederic C Ludwig and Robert M Elashoff
1972 Transactions of The New York Academy of Sciences 34(7): 582-587.

59. **The ageing systemic milieu negatively regulates neurogenesis and cognitive function**
Saul A Villeda, Jian Luo, et al.
2011 Nature 477: 90-94.

60. **Young blood reverses age-related impairments in cognitive function and synaptic plasticity in mice**
Saul A Villeda, Kristopher E Plambeck, et al.
2014 Nature Medicine 20: 659-663.

61. **Studies that shed new light on aging**
Harold L Katcher
2013 Biochemistry Moscow 38: 1061-70.

62. **Towards an evidence-based model of aging**
Harold L Katcher
2015 Current Aging Science 8(1): 46-55.

63. **Plasma dilution improves cognition and attenuates neuroinflammation in old mice**
Melod Mehdipour, Taha Mehdipour, et al.
2021 GeroScience 43: 1–18.

64. **Human umbilical cord plasma proteins revitalize hippocampal function in aged mice**
Joseph M Castellano, Kira I Mosher, et al.
2017 Nature 544: 488–492.

65. **Universal DNA methylation age across mammalian tissues**
Mammalian Methylation Consortium: Ake T Lu, Zhe Fei, Amin Haghani, et al.
2021 bioRxiv 2021.01.18.426733.

66. **Notch-mediated restoration of regenerative potential to aged muscle**
Irina M Conboy, Michael J Conboy, et al.
2003 Science 302(5650): 1575-1577.

67. **Studies Financed by Heales: Effect of Young Rat Plasma on The Lifespan of Aging Rats**
Available at https://heales.org/2020/12/22/studies-financed-by-heales-effect-of-young-rat-plasma-on-the-lifespan-of-aging-rats-21-december-2020/

68. **Rejuvenating senescent and centenarian human cells by reprogramming through the pluripotent state**
Laure Lapasset, Ollivier Milhavet, et al.
2011 Genes & Development 25(21): 2248–2253.

69. **In Contrast to Dolly, Cloning Resets Telomere Clock in Cattle**
Gretchen Vogel
2000 Science 288(5466): 586-587.

70. **Reprogramming to recover youthful epigenetic information and restore vision**
Yuancheng Lu, Benedikt Brommer, et al.
2020 Nature 588:124-129.

71. **In Vivo Amelioration of Age-Associated Hallmarks by Partial Reprogramming**
Alejandro Ocampo, Pradeep Reddy, et al.
2016 Cell 167: 1719–1733.

72. **Turning back time with emerging rejuvenation strategies**
Salah Mahmoudi, Lucy Xu and Anne Brunet
2019 Nature Cell Biology 21: 32–43.

73. **The effect of retarded growth upon the length of life span and upon the ultimate body size**
C M McCay, M F Crowell and L A Maynard
1935 The Journal of Nutrition 10(1): 63–79.

74. **Clearance of p16^{Ink4a}-positive senescent cells delays ageing-associated disorders**
Darren J. Baker, Tobias Wijshake, et al.
2011 Nature volume 479: 232–236.

75. **What is Life? With Mind and Matter and Autobiographical Sketches**
Erwin Schrödinger
2012 Cambridge University Press; Reprint edition

76. **The Biological Clock in Gray Mouse Lemur: Adaptive, Evolutionary and Aging Considerations in an Emerging Non-human Primate Model**
Clara Hozer, Fabien Pifferi, Fabienne Aujard and Martine Perret
2019 Frontiers in Physiology 10, article 1033.

Made in the USA
Middletown, DE
30 August 2021